Topics in Recreational Mathematics 4/2015

Editor-in-chief

Charles Ashbacher
5530 Kacena Ave
Marion, IA 52302 USA

cashbacher@yahoo.com

Assistant Editors

Rachel Pollari

Jennifer Corrigan

Artwork

Caytie Ribble

Problems

Lamarr Widmer

ISBN: 978-1514317518

CONTENTS

NOTE FROM THE EDITOR

Welcome to the fourth edition of "Topics in Recreational Mathematics." This one starts off with three mathematical cartoons by Caytie Ribble, they will appear in an upcoming book of mathematical cartoons and humor that will be published in the middle months of 2015. At the end there is a page devoted to each of the books that has been published by Charles Ashbacher and his associates.

An analysis of the relationship between birth month and hockey performance, magic squares and cubes, concatenation sequences and a short biography are some of the other items that are included. The biography of al-Kashi is the first of what is to be a series of items about Arab and Persian mathematical achievements. Alphametics, problems and book reviews are included as well.

As always, I thank you for your support and welcome your comments and input at

cashbacher@yahoo.com

Contributions are also actively solicited and should be sent to the previous electronic address.

Charles Ashbacher

A normal curve is perfectly
balanced about the mean

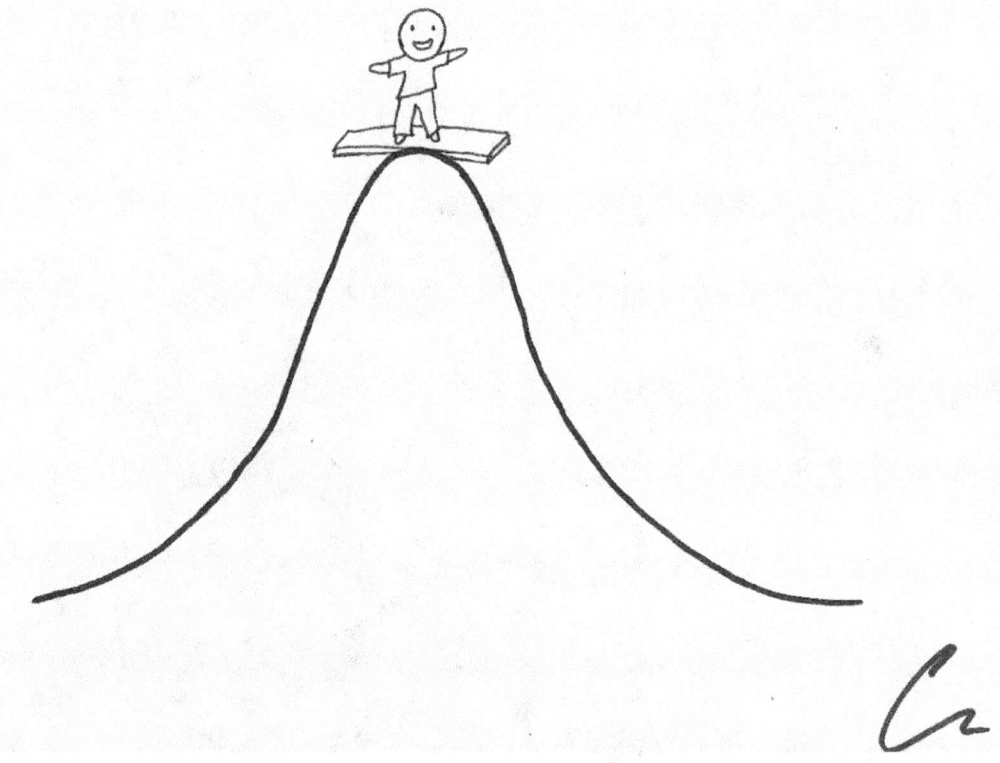

HIPPOPEDE CURVE

ARE CANADIAN NHL HALL OF FAMERS WINTER BABIES?

Arthur E. Mittnacht

Paul M. Sommers

Department of Economics

Middlebury College

Middlebury, Vermont 05753

psommers@middlebury.edu

Abstract

Canadian journalist Malcolm Gladwell has suggested that hockey players (particularly Canadians) with birthdays early in the year have a greater chance of becoming elite players than those with birthdays late in the year. This note examines all Canadian players (by position, by birthdate, and by province of birth) elected to the NHL Hall of Fame through the year 2008. The first three months in the Gladwell division are January, February, and March; the first three months in the seasonal division are December, January, and February. Chi-square goodness-of-fit tests show no empirical support for the Gladwell breakdown, but disproportionately many winter babies for the seasonal breakdown.

Outliers: The Story of Success, by Malcolm Gladwell, suggests that hockey players (particularly Canadians) with birthdays in earlier months have a greater chance of becoming elite players than those with birthdays in later months [1]. After all, if the eligibility cutoff for age-class hockey is January 1st, then a boy born in January has almost an entire extra year to grow than another boy born in December, yet they both must compete for the same roster spot.[1] Gladwell [1, pp. 20-21], for example, notes the disproportionately large number of young men (almost half of the entire roster) born in January, February, and March on the 2007 Medicine Hat Tigers Canadian Junior A league hockey team.

The purpose of this brief note is to examine the birth dates of hockey's elite, Canadian players elected to the National Hockey League (NHL) Hall of Fame through the year 2008. The data on birth dates are from www.legendsofhockey.net/html/legendsplayer.htm .

The calendar year will be divided two different ways. First, the calendar year will be divided into three-month periods (hereafter, the Gladwell division): January, February, March in the first period (hereafter, the first quarter); April, May, June in the second (hereafter, the second quarter); and so forth. Alternatively, one might want to test the belief that the elite players in a winter sport were generally born in winter months (henceforth, the seasonal division). The winter months are here assumed to be December, January, and February; the spring months are March, April, and May; the summer months are June, July, and August; and the fall months are September, October, and November. Of the 240 players inducted into the NHL's Hall of Fame through the year 2008, 219 were born in Canada (with birth dates reported for 213 of these inductees). And, of the 213 Canadian players enshrined in the Hall of Fame, 29.1 percent were born in the first quarter, 22.1 percent were born in the second quarter, 22.5 percent were born in the third, and 26.3 percent were born in the fourth. By comparison, 32.9 percent were born in the winter, 22.1 percent were born in the spring, 23.9 percent were born in the summer, and 21.1 percent were born in the fall.

Table 1 shows the Gladwell breakdown of births for various subgroups of Canadian born inductees. Table 2 shows a similar breakdown by season. To determine whether there are statistically discernible variations, either by quarter or by season, a χ^2 goodness-of-fit test was done of the null hypothesis that births are spread evenly over all four three-month periods (quarterly or by season).

In view of the χ^2 values reported in the last column of Tables 1 and 2, the null hypothesis cannot be rejected (at better than the .05 significance level) using Gladwell's quarterly division.[2] But, the null hypothesis *can* be rejected in several instances using the seasonal breakdown. And, in particular, Canadian Hall of Famers, notably defensemen, born in Quebec or before 1944 were winter babies more often than not.

Table 1. Births by Quarter for Various Groups of

Canadian NHL Hall of Famers

Group	Quarter				
	First	Second	Third	Fourth	χ^2
All inductees	62	47	48	56	2.831
Position					
Forwards[a]	32	19	30	28	3.624
Defensemen[b]	20	14	10	11	4.418
Birthdate					
Before 1944[c]	52	38	36	47	3.948
After 1944	10	9	12	9	0.600
Province[d]					
Ontario	32	25	25	30	1.357
Quebec	20	8	9	15	7.231***

*Significant at better than the .01 level.

**Significant at better than the .05 level.

***Significant at better than the .10 level.

[a] Forwards include right wing, center, and left wing.

[b] The difference between "All inductees" and the sum of "Forwards" and "Defensemen" include goalies, rovers, and players identified as playing multiple positions.

[c] No inductees were born in 1944.

[d] Canadian Hall of Famers include natives of other provinces: Manitoba (18), Saskatchewan (17), Alberta (8), British Columbia (4), and one each from New Brunswick, Nova Scotia, and Newfoundland.

Table 2. Births by Season for Various Groups of

Canadian NHL Hall of Famers

	Season				
Group	Winter	Spring	Summer	Fall	χ^2
All inductees	70	47	51	45	7.376**
Position					
Forwards[a]	35	19	28	27	4.725
Defensemen[b]	22	16	13	4	12.273*
Birthdate					
Before 1944[c]	61	37	40	35	10.006**
After 1944	9	10	11	10	0.200
Province[d]					
Ontario	32	25	25	30	1.357
Quebec	20	8	9	15	7.231***

*Significant at better than the .01 level.

**Significant at better than the .05 level.

***Significant at better than the .10 level.

[a] See footnote *a* in Table 1.

[b] See footnote *b* in Table 1.

[c] See footnote *c* in Table 1.

[d] See footnote *d* in Table 1.

Reference

1.   M. Gladwell, *Outliers: The Story of Success*, Little, Brown and Co., New York, NY, 2008.

Footnotes

1.   Two notable January babies now enshrined in the NHL Hall of Fame are Wayne Gretzky (born on January 26, 1961) and Mark Messier (born on January 18, 1961).

2.   We cannot reject the null hypothesis of no difference in the birth rate between months ($\chi^2 = 9.366$, $p = .588$). There are as many Canadian Hall of Famers born in January as there are in December (25).

CONCENTRIC MAGIC CUBES OF PRIME NUMBERS

Natalia Makarova

Saratov, Russia

natalimak1@yandex.ru

Abstract

Like their two-dimensional counterparts, three-dimensional magic cubes can fascinate and surprise you with their existence. The level of difficulty and fascination is even higher when magic cubes are constructed inside magic cubes. When there are several layers, they remind you of the Russian Matryoshka dolls-within-dolls. Several examples of magic cubes constructed using formulas are given in this paper, including some constructed from prime numbers.

Introduction

On the web page

http://www.magic-SquareS.net/c-t-htm/c_prime.htm

there is the concentric magic cube of order 6 of distinct primes seen in figure 1.

Figure 1

4831	4783	67	9811	4639	5479
191	241	193	9473	9769	9743
331	577	5009	4751	9619	9323
8273	9719	8933	1123	829	733
8423	7499	8287	1789	1801	1811
7561	6791	7121	2663	2953	2521

131	761	379	9403	9497	9439
8951	2437	3547	5309	8447	919
9643	3209	5573	2281	8677	227
2143	8243	4877	6007	613	7727
8311	5851	5743	6143	2003	1559
431	9109	9491	467	373	9739

337	8849	8821	1409	1307	8887
7013	5903	2879	9007	1951	2857
8009	3217	2767	8117	5639	1861
9049	6073	5521	2333	5813	821
4219	4547	8573	283	6337	5651
983	1021	1049	8461	8563	9533

8543	8839	9277	173	1831	947
4177	3533	9587	1297	5323	5693
7487	4057	7537	4349	3797	2383
31	4231	1753	7103	6653	9839
449	7919	863	6991	3967	9421
8923	1031	593	9697	8039	1327

8419	3299	8317	1607	5419	2549
9151	7867	3727	4127	4019	719
3593	9257	3863	4993	1627	6277
977	1193	7589	4297	6661	8893
149	1423	4561	6323	7433	9721
7321	6571	1553	8263	4451	1451

7349	3079	2749	7207	6917	2309
127	9629	9677	397	101	9679
547	9293	4861	5119	251	9539
9137	151	937	8747	9041	1597
8059	2371	1583	8081	8069	1447
4391	5087	9803	59	5231	5039

The magic constant of this magic cube is 29610.

Inside is the associative (central symmetric) and pantriagonal cube of order 4 with a magic constant of 19740 that is seen in figure 2.

Figure 2

2437	3547	5309	8447
3209	5573	2281	8677
8243	4877	6007	613
5851	5743	6143	2003

5903	2879	9007	1951
3217	2767	8117	5639
6073	5521	2333	5813
4547	8573	283	6337

3533	9587	1297	5323
4057	7537	4349	3797
4231	1753	7103	6653
7919	863	6991	3967

7867	3727	4127	4019
9257	3863	4993	1627
1193	7589	4297	6661
1423	4561	6323	7433

The concentric magic cube is an interesting subset of the set of magic cubes.

Our next step will be to present concentric magic cubes of orders 5, 6 and 7 that are composed of distinct primes.

Concentric magic cubes of order 5

The scheme for concentric magic cubes of order 5 is as seen in figure 3

Figure 3

x1	x2	x3	x4	x5
x6	x7	x8	x9	x10
x11	x12	x13	x14	x15
x16	x17	x18	x19	x20
x21	x22	x23	x24	x25

x26	x27	x28	x29	x30
x31				k-x31
x32				k-x32
x33				k-x33
k-x30	k-x27	k-x28	k-x29	k-x26

x34	x35	x36	x37	x38
x39				k-x39
x40		k/2		k-x40
x41				k-x41
k-x38	k-x35	k-x36	k-x37	k-x34

x42	x43	x44	x45	x46
x47				k-x47
x48				k-x48
x49				k-x49
k-x46	k-x43	k-x44	k-x45	k-x42

k-x25	k-x22	k-x23	k-x24	k-x21
k-x10	k-x7	k-x8	k-x9	k-x6
k-x15	k-x12	k-x13	k-x14	k-x11
k-x20	k-x17	k-x18	k-x19	k-x16
k-x5	k-x2	k-x3	k-x4	k-x1

The magic constant of this magic cube $S = 5k / 2$.

In the inner region we have the magic cube of order 3 with magic constant $S = 3k / 2$. This magic cube is associative with constant associativity k.

We make this magic cube in the first stage.

If such a magic cube of order 3 is found, the second stage is the working out of the edging.

The system of linear equations describing the concentric cube of order 5 follows.

$x1+x2+x3+x4+x5=S$

$x6+x7+x8+x9+x10=S$

$x11+x12+x13+x14+x15=S$

$x16+x17+x18+x19+x20=S$

$x21+x22+x23+x24+x25=S$

$x26+x27+x28+x29+x30=S$

$x34+x35+x36+x37+x38=S$

$x42+x43+x44+x45+x46=S$

$$x1+x6+x11+x16+x21=S$$

$$x2+x7+x12+x17+x22=S$$

$$x3+x8+x13+x18+x23=S$$

$$x4+x9+x14+x19+x24=S$$

$$x5+x10+x15+x20+x25=S$$

$$x26+x31+x32+x33-x30=3S/5$$

$$x34+x39+x40+x41-x38=3S/5$$

$$x42+x47+x48+x49-x46=3S/5$$

$$x1+x26+x34+x42-x25=3S/5$$

$$x2+x27+x35+x43-x22=3S/5$$

$$x3+x28+x36+x44-x23=3S/5$$

$$x4+x29+x37+x45-x24=3S/5$$

$$x5+x30+x38+x46-x21=3S/5$$

$$x6+x31+x39+x47-x10=3S/5$$

$$x11+x32+x40+x48-x15=3S/5$$

$$x16+x33+x41+x49-x20=3S/5$$

This system is solved to find the general formula for the concentric magic cube of order 5.

$$x5=S-x1-x2-x3-x4$$
$$x9=S-x10-x6-x7-x8$$
$$x15=S-x11-x12-x13-x14$$
$$x17=S-x16-x18-x19-x20$$
$$x21=S-x1-x11-x16-x6$$
$$x22=-x12+x16+x18+x19-x2+x20-x7$$
$$x23=S-x13-x18-x3-x8$$
$$x24=x10-x14-x19-x4+x6+x7+x8$$
$$x25=-S+x1-x10+x11+x12+x13+x14+x2-x20+x3+x4$$
$$x30=S-x26-x27-x28-x29$$
$$x33=(8\ S)/5-2\ x26-x27-x28-x29-x31-x32$$
$$x34=-S-x10+x11+x12+x13+x14+x2-x20-x26+x3+x4-x46+x47+x48+x49$$
$$x35=S/5-x12+x16+x18+x19-2\ x2+x20-x27+x44+x45+2\ x46-x47-x48-x49-x7$$

x36=(8 S)/5-x13-x18-x28-2 x3-x44-x8

x37=(3 S)/5+x10-x14-x19-x29-2 x4-x45+x6+x7+x8

x38=-((2 S)/5)-x11-x16+x2+x26+x27+x28+x29+x3+x4-x46-x6

x39=(3 S)/5+x10-x31-x47-x6

x40=(8 S)/5-2 x11-x12-x13-x14-x32-x48

x41=-S-x16+x20+2 x26+x27+x28+x29+x31+x32-x49

x42=(3 S)/5+x46-x47-x48-x49

x43=(2 S)/5-x44-x45-2 x46+x47+x48+x49

This system as 28 of the 49 variables free once you set the parameter k.

Using this general formula, you can construct a lot of concentric magic cubes of order 5.

Here are some of my solutions for specific values of S and k.

Figure 4

S = 12955, k = 5182

4253	953	1301	2789	3659
701	2909	4133	4721	491
431	2099	2633	5051	2741
3989	3371	809	83	4703
3581	3623	4079	311	1361

29	2063	4799	3923	2141
2213	2939	4733	101	2969
4001	3863	659	3251	1181
3671	971	2381	4421	1511
3041	3119	383	1259	5153

173	3449	4271	599	4463
5171	4073	239	3461	11
4919	1979	2591	3203	263
1973	1721	4943	1109	3209
719	1733	911	4583	5009

4679	4931	1481	773	1091
179	761	2801	4211	5003
1163	1931	4523	1319	4019
2843	5081	449	2243	2339
4091	251	3701	4409	503

3821	1559	1103	4871	1601
4691	2273	1049	461	4481
2441	3083	2549	131	4751
479	1811	4373	5099	1193
1523	4229	3881	2393	929

Figure 5

S = 13945, k = 5578

2111	2039	5399	107	4289
5519	1187	2741	4271	227
557	4481	389	3221	5297
4817	1229	269	4139	3491
941	5009	5147	2207	641

1559	5507	1997	4877	5
857	2411	4259	1697	4721
5477	2957	2441	2969	101
479	2999	1667	3701	5099
5573	71	3581	701	4019

1571	4649	1907	4931	887
1901	4079	197	4091	3677
4751	2801	2789	2777	827
1031	1487	5381	1499	4547
4691	929	3671	647	4007

3767	1181	4211	659	4127
317	1877	3911	2579	5261
2879	2609	3137	2621	2699
5531	3881	1319	3167	47
1451	4397	1367	4919	1811

4937	569	431	3371	4637
5351	4391	2837	1307	59
281	1097	5189	2357	5021
2087	4349	5309	1439	761
1289	3539	179	5471	3467

Figure 6

S = 15835, k = 6334

3923	4337	941	5273	1361
431	6329	3533	71	5471
5693	1031	4457	4463	191
5507	4127	2633	857	2711
281	11	4271	5171	6101

683	4091	3557	5591	1913
5237	3491	5657	353	1097
947	3593	761	5147	5387
4547	2417	3083	4001	1787
4421	2243	2777	743	5651

5153	467	2963	3671	3581
5051	3677	593	5231	1283
2531	4721	3167	1613	3803
347	1103	5741	2657	5987
2753	5867	3371	2663	1181

5843	617	6311	137	2927
4253	2333	3251	3917	2081
521	1187	5573	2741	5813
1811	5981	677	2843	4523
3407	5717	23	6197	491

233	6323	2063	1163	6053
863	5	2801	6263	5903
6143	5303	1877	1871	641
3623	2207	3701	5477	827
4973	1997	5393	1061	2411

Figure 7

S = 15955, k = 6382

4001	5099	1439	3023	2393
479	6143	1901	1103	6329
5471	641	3923	5639	281
5981	353	2579	4889	2153
23	3719	6113	1301	4799

83	1709	6131	5189	2843
5441	3371	5153	1049	941
2243	3413	1889	4271	4139
4649	2789	2531	4253	1733
3539	4673	251	1193	6299

4409	4973	3833	2141	599
3671	4073	569	4931	2711
1109	4049	3191	2333	5273
983	1451	5813	2309	5399
5783	1409	2549	4241	1973

5879	1511	4283	521	3761
6311	2129	3851	3593	71
1031	2111	4493	2969	5351
113	5333	1229	3011	6269
2621	4871	2099	5861	503

1583	2663	269	5081	6359
53	239	4481	5279	5903
6101	5741	2459	743	911
4229	6029	3803	1493	401
3989	1283	4943	3359	2381

Figure 8

S = 18035, k = 7214

4297	5647	3571	157	4363
2707	967	2617	6841	4903
2887	4933	5791	2953	1471
1867	2971	3673	6793	2731
6277	3517	2383	1291	4567

523	6427	877	5197	5011
6481	3943	6571	307	733
5437	4657	1063	5101	1777
3391	2221	3187	5413	3823
2203	787	6337	2017	6691

4447	331	6163	541	6553
5683	5077	223	5521	1531
607	4051	3607	3163	6607
6637	1693	6991	2137	577
661	6883	1051	6673	2767

6121	1933	2593	6217	1171
853	1801	4027	4993	6361
3361	2113	6151	2557	3853
1657	6907	643	3271	5557
6043	5281	4621	997	1093

2647	3697	4831	5923	937
2311	6247	4597	373	4507
5743	2281	1423	4261	4327
4483	4243	3541	421	5347
2851	1567	3643	7057	2917

In figures 25, 27 and 28 you can see interior concentric magic cubes of order 5 inside concentric magic cubes of order 7.

It is interesting to note that we can create classic (numbers 1 though n^2) concentric magic cubes of order 5 (see Fig. 9). This magic cube is made up of distinct positive integers from 1 to 125.

Figure 9

S = 315, k = 126

16	98	62	25	114
44	75	112	50	34
43	74	49	113	36
115	41	73	48	38
97	27	19	79	93

21	10	70	125	89
94	81	106	2	32
91	100	22	67	35
72	8	61	120	54
37	116	56	1	105

122	68	5	103	17
46	102	18	69	80
7	30	63	96	119
31	57	108	24	95
109	58	121	23	4

28

123	40	71	15	66
39	6	65	118	87
84	59	104	26	42
9	124	20	45	117
60	86	55	111	3

33	99	107	47	29
92	51	14	76	82
90	52	77	13	83
88	85	53	78	11
12	28	64	101	110

Concentric magic cubes of order 6

A concentric magic cube of order 6 has already appeared in figure 1. A scheme for concentric magic cubes of order 6 is given in figure 10.

Figure 10

x1	x2	x3	x4	x5	x6
x7	x8	x9	x10	x11	x12
x13	x14	x15	x16	x17	x18
x19	x20	x21	x22	x23	x24
x25	x26	x27	x28	x29	x30
x31	x32	x33	x34	x35	x36

29

x37	x38	x39	x40	x41	x42
x43					k-x43
x44					k-x44
x45					k-x45
x46					k-x46
k-x42	k-x38	k-x39	k-x40	k-x41	k-x37

x47	x48	x49	x50	x51	x52
x53					k-x53
x54					k-x54
x55					k-x55
x56					k-x56
k-x52	k-x48	k-x49	k-x50	k-x51	k-x47

x57	x58	x59	x60	x61	x62
x63					k-x63
x64					k-x64
x65					k-x65
x66					k-x66
k-x62	k-x58	k-x59	k-x60	k-x61	k-x57

x67	x68	x69	x70	x71	x72
x73					k-x73
x74					k-x74
x75					k-x75
x76					k-x76
k-x72	k-x68	k-x69	k-x70	k-x71	k-x67

k-x36	k-x32	k-x33	k-x34	k-x35	k-x31
k-x12	k-x8	k-x9	k-x10	k-x11	k-x7
k-x18	k-x14	k-x15	k-x16	k-x17	k-x13
k-x24	k-x20	k-x21	k-x22	k-x23	k-x19
k-x30	k-x26	k-x27	k-x28	k-x29	k-x25

The magic constant of the cube $S = 3k$. Inside there is a magic cube of order 4 with a magic constant $S = 2k$, this magic cube can be both associative and non-associative. In magic cube of figure 1 is an example of an interior magic cube that is associative.

In the present scheme one can write a system of linear equations and solve it. The end result is a general formula for concentric magic cubes of order 6, similar to that shown for the concentric magic cube of order 5.

Using distinct primes I was able to create several concentric magic cubes of order 6 with a magic constant $S < 29610$. Here are some examples.

Figure 11

S = 5670, k = 1890

971	761	1801	367	157	1613
1447	379	491	1031	569	1753
1033	1667	1709	281	877	103
1117	1777	419	457	1607	293
461	23	787	1723	797	1879
641	1063	463	1811	1663	29

607	409	719	1877	1627	431
1543	1039	31	887	1823	347
599	439	2311	383	647	1291
233	353	557	2371	499	1657
1229	1949	881	139	811	661
1459	1481	1171	13	263	1283

937	149	1303	677	983	1621
1553	97	307	2069	1307	337
727	2267	523	317	673	1163
653	967	1451	271	1091	1237
1531	449	1499	1123	709	359
269	1741	587	1213	907	953

1277	1693	19	1289	829	563
593	53	2333	751	643	1297
1423	991	107	373	2309	467
311	2423	571	659	127	1579
739	313	769	1997	701	1151
1327	197	1871	601	1061	613

17	1831	401	1381	1847	193
397	2591	1109	73	7	1493
101	83	839	2707	151	1789
1759	37	1201	479	2063	131
1699	1069	631	521	1559	191
1697	59	1489	509	43	1873

1861	827	1427	79	227	1249
137	1511	1399	859	1321	443
1787	223	181	1609	1013	857
1597	113	1471	1433	283	773
11	1867	1103	167	1093	1429
277	1129	89	1523	1733	919

This magic cube is a known minimal simple magic cube.

Figure 12

S = 6030, k = 2010

13	59	1319	1439	1987	1213
1811	479	397	257	1759	1327
1973	383	839	1277	1091	467
1103	1949	1283	461	353	881
991	1259	499	1409	439	1433
139	1901	1693	1187	401	709

7	389	1777	1487	1913	457
1033	1861	19	641	1499	977
643	877	1597	1373	173	1367
1847	1193	71	1459	1297	163
947	89	2333	547	1051	1063
1553	1621	233	523	97	2003

809	1601	1291	557	31	1741
1567	151	73	1667	2129	443
1151	953	1723	587	757	859
787	283	1733	1303	701	1223
1447	2633	491	463	433	563
269	409	719	1453	1979	1201

34

1993	1999	1013	487	311	227
79	449	1889	1471	211	1931
347	1279	107	2017	617	1663
941	1061	1867	431	661	1069
887	1231	157	101	2531	1123
1783	11	997	1523	1699	17

1907	1873	313	1237	179	521
857	1559	2039	241	181	1153
373	911	593	43	2473	1637
223	1483	349	827	1361	1787
1181	67	1039	2909	5	829
1489	137	1697	773	1831	103

1301	109	317	823	1609	1871
683	1531	1613	1753	251	199
1543	1627	1171	733	919	37
1129	61	727	1549	1657	907
577	751	1511	601	1571	1019
797	1951	691	571	23	1997

Figure 13

S = 10080, k = 3360

421	2347	2081	2971	1583	677
1277	1229	1307	3323	1571	1373
1549	1933	2477	293	1297	2531
2539	359	2069	2423	1951	739
2393	1861	653	499	2297	2377
1901	2351	1493	571	1381	2383

1481	41	197	2719	2521	3121
673	919	3331	1913	557	2687
1873	1663	151	2273	2633	1487
2851	3329	17	193	3181	509
2963	809	3221	2341	349	397
239	3319	3163	641	839	1879

1223	463	3079	857	2207	2251
2837	661	2269	3347	443	523
2903	3251	103	53	3313	457
647	2617	2309	61	1733	2713
1361	191	2039	3259	1231	1999
1109	2897	281	2503	1153	2137

2677	3083	2767	311	541	701
1693	2129	101	1321	3169	1667
1367	1627	3299	1051	743	1993
1181	47	3307	3257	109	2179
503	2917	13	1091	2699	2857
2659	277	593	3049	2819	683

3301	3137	89	433	1249	1871
1613	3011	1019	139	2551	1747
1559	179	3167	3343	31	1801
241	727	1087	3209	1697	3119
1877	2803	1447	29	2441	1483
1489	223	3271	2927	2111	59

977	1009	1867	2789	1979	1459
1987	2131	2053	37	1789	2083
829	1427	883	3067	2063	1811
2621	3001	1291	937	1409	821
983	1499	2707	2861	1063	967
2683	1013	1279	389	1777	2939

Figure 14

S = 19800, k = 6600

883	5591	3697	4243	233	5153
1117	4457	5923	1031	4799	2473
5387	4289	977	2273	5507	1367
4447	179	2861	6121	743	5449
6269	863	5323	1549	2777	3019
1697	4421	1019	4583	5741	2339

1783	47	239	6373	6007	5351
151	613	6571	4349	1667	6449
6247	3517	283	5147	4253	353
6211	6359	23	271	6547	389
4159	2711	6323	3433	733	2441
1249	6553	6361	227	593	4817

397	211	3299	5861	5981	4051
4597	1321	3391	6491	1997	2003
1523	6563	19	149	6469	5077
6221	4363	5273	31	3533	379
4513	953	4517	6529	1201	2087
2549	6389	3301	739	619	6203

6133	6113	3797	449	3229	79
3881	5399	71	2083	5647	2719
911	3067	6569	1327	2237	5689
1597	131	6451	6581	37	5003
757	4603	109	3209	5279	5843
6521	487	2803	6151	3371	467

6343	5659	3187	857	3491	263
5927	5867	3167	277	3889	673
499	53	6329	6577	241	6101
173	2347	1453	6317	3083	6427
521	4933	2251	29	5987	6079
6337	941	3413	5743	3109	257

4261	2179	5581	2017	859	4903
4127	2143	677	5569	1801	5483
5233	2311	5623	4327	1093	1213
1151	6421	3739	479	5857	2153
3581	5737	1277	5051	3823	331
1447	1009	2903	2357	6367	5717

Figure 15

S = 25200, k = 8400

5197	5521	5171	4373	677	4261
2311	2699	7237	1361	6029	5563
6803	919	4787	8053	311	4327
5087	7937	1543	1747	7433	1453
571	2267	3413	3469	8171	7309
5231	5857	3049	6197	2579	2287

523	809	11	8269	8011	7577
37	1213	8293	2621	4673	8363
8161	7573	643	5807	2777	239
8287	6911	137	2011	7741	113
7369	1103	7727	6361	1609	1031
823	7591	8389	131	389	7877

179	6763	5717	3389	5689	3463
8317	709	6397	6287	3407	83
2503	8387	1399	2153	4861	5897
2143	727	7823	31	8219	6257
7121	6977	1181	8329	313	1279
4937	1637	2683	5011	2711	8221

4951	7537	1667	1567	4241	5237
7901	8087	71	7219	1423	499
1187	181	8369	577	7673	7213
1297	3539	6247	7001	13	7103
6701	4993	2113	2003	7691	1699
3163	863	6733	6833	4159	3449

8237	2027	7283	5399	761	1493
3797	6791	2039	673	7297	4603
2473	659	6389	8263	1489	5927
1439	5623	2593	7757	827	6961
2347	3727	5779	107	7187	6053
6907	6373	1117	3001	7639	163

6113	2543	5351	2203	5821	3169
2837	5701	1163	7039	2371	6089
4073	7481	3613	347	8089	1597
6947	463	6857	6653	967	3313
1091	6133	4987	4931	229	7829
4139	2879	3229	4027	7723	3203

I was unable to make a concentric magic cube of order 6 with a magic constant S = 5040, perhaps a reader will have better luck.

The border is easily made. One embodiment is shown in Fig. 16.

Figure 16

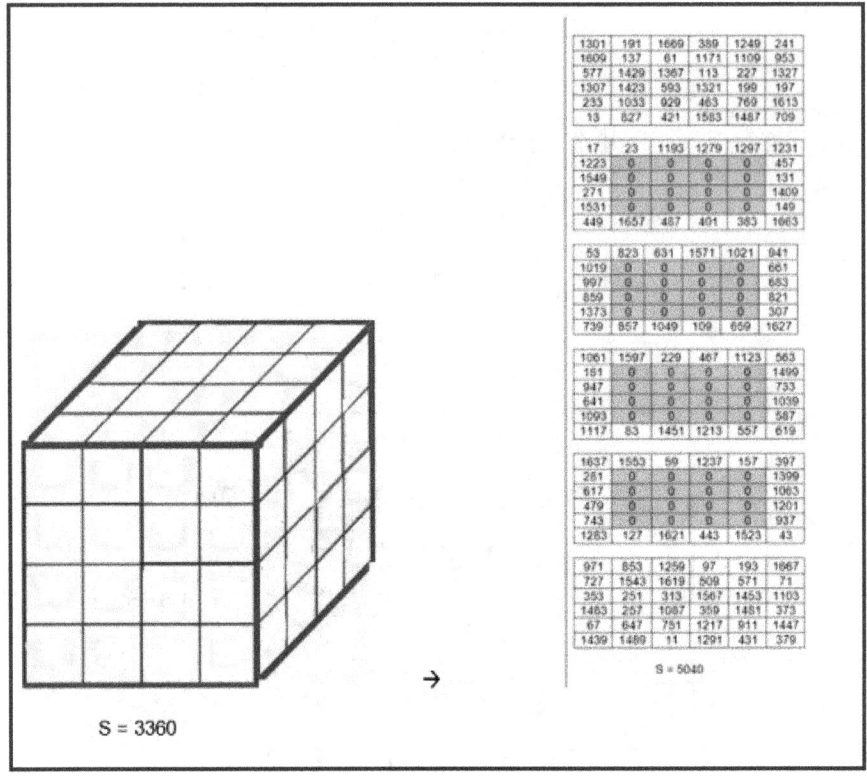

Try to find a magic cube of order 4 with a magic constant S = 3360, which is to be inserted into the fringes, shown on the right in the illustration. It's an interesting challenge!

My partial solution to the problem that contains 5 errors appears in figure 17.

Figure 17

1301	191	1669	389	1249	241
1609	137	61	1171	1109	953
577	1429	1367	113	227	1327
1307	1423	593	1321	199	197
233	1033	929	463	769	1613
13	827	421	1583	1487	709

42

17	23	1193	1279	1297	1231
1223	1987	7	863	503	457
1549	613	2221	107	419	131
271	239	101	2053	967	1409
1531	521	1031	337	1471	149
449	1657	487	401	383	1663

53	823	631	1571	1021	941
1019	103	73	1997	1187	661
997	1847	673	317	523	683
859	601	461	1009	1289	821
1373	809	2153	37	361	307
739	857	1049	109	659	1627

1061	1597	229	467	1123	563
181	701	2099	349	211	1499
947	409	311	919	1721	733
641	1973	787	269	331	1039
1093	277	163	1823	1097	587
1117	83	1451	1213	557	619

1637	1553	59	1237	157	397
281	569	1181	151	1459	1399
617	491	155	2017	697	1063
479	547	2011	29	773	1201
743	1753	13	1163	431	937
1283	127	1621	443	1523	43

971	853	1259	97	193	1667
727	1543	1619	509	571	71
353	251	313	1567	1453	1103
1483	257	1087	359	1481	373
67	647	751	1217	911	1447
1439	1489	11	1291	431	379

Explanation of the errors:

361 is not prime
155 is not prime
697 is not prime
13 is not unique
431 is not unique

Another embodiment of the border appears in figure 18.

Figure 18

1327	739	479	821	1663	11
1453	149	13	1367	1087	971
1307	1093	1637	647	127	229
181	1291	1283	769	257	1259
641	1301	307	1237	383	1171
131	467	1321	199	1523	1399

23	53	1567	1021	1429	947
883					797
857					823
1483					197
1061					619
733	1627	113	659	251	1657

191	109	1063	853	1223	1601
937					743
631					1049
1619					61
1583					97
79	1571	617	827	457	1489

1597	1439	449	463	431	661
571					1109
67					1613
1103					577
683					997
1019	241	1231	1217	1249	83

1621	1487	1123	401	137	271
487					1193
727					953
233					1447
563					1117
1409	193	557	1279	1543	59

281	1213	359	1481	157	1549
709	1531	1667	313	593	227
1451	587	43	1033	1553	373
421	389	397	911	1423	1499
509	379	1373	443	1297	1039
1669	941	1201	859	17	353

I found a solution for this variant that only has three errors and it appears in figure 19.

Explanation of the errors:

445 is not prime

781 is not prime

97 is not unique

Figure 19

S = 5040, k = 1680

1327	739	479	821	1663	11
1453	149	13	1367	1087	971
1307	1093	1637	647	127	229
181	1291	1283	769	257	1259
641	1301	307	1237	383	1171
131	467	1321	199	1523	1399

23	53	1567	1021	1429	947
883	1987	7	863	503	797
857	613	2221	107	419	823
1483	239	101	2053	967	197
1061	521	1031	337	1471	619
733	1627	113	659	251	1657

191	109	1063	853	1223	1601
937	103	73	1997	1187	743
631	1847	757	317	439	1049
1619	601	461	1009	1289	61
1583	809	2069	37	445	97
79	1571	617	827	457	1489

1597	1439	449	463	431	661
571	701	2099	349	211	1109
67	409	311	919	1721	1613
1103	1973	787	269	331	577
683	277	163	1823	1097	997
1019	241	1231	1217	1249	83

1621	1487	1123	401	137	271
487	569	1181	151	1459	1193
727	491	71	2017	781	953
233	547	2011	29	773	1447
563	1753	97	1163	347	1117
1409	193	557	1279	1543	59

281	1213	359	1481	157	1549
709	1531	1667	313	593	227
1451	587	43	1033	1553	373
421	389	397	911	1423	1499
509	379	1373	443	1297	1039
1669	941	1201	859	17	353

Figure 20 contains a classic concentric magic cube of order 6. This cube is made up of distinct positive integers from 1 to 216. The interior is an unconventional associative magic cube of order 4 with magic constant S = 434.

Figure 20

S = 651, k = 217.

107	160	56	78	133	117
104	165	51	76	134	121
193	155	70	81	37	115
9	153	87	146	179	77
188	17	214	181	16	35
50	1	173	89	152	186

75	48	82	42	194	210
156	2	99	144	189	61
162	79	46	197	112	55
148	150	183	8	93	69
103	203	106	85	40	114
7	169	135	175	23	142

108	12	45	192	174	120
74	98	27	164	145	143
129	47	54	137	196	88
127	158	151	92	33	90
116	131	202	41	60	101
97	205	172	25	43	109

123	30	211	206	63	18
95	157	176	15	86	122
26	184	125	66	59	191
159	21	80	163	170	58
49	72	53	190	119	168
199	187	6	11	154	94

207	185	213	5	22	19
126	177	132	111	14	91
39	124	209	34	67	178
68	105	20	171	138	149
13	28	73	118	215	204
198	32	4	212	195	10

31	216	44	128	65	167
96	52	166	141	83	113
102	62	147	136	180	24
140	64	130	71	38	208
182	200	3	36	201	29
100	57	161	139	84	110

Concentric magic cubes of order 7

Figure 21 contains a scheme for concentric magic cubes of order 7.

Figure 21

y1	y2	y3	y4	y5	y6	y7
y8	y9	y10	y11	y12	y13	y14
y15	y16	y17	y18	y19	y20	y21
y22	y23	y24	y25	y26	y27	y28
y29	y30	y31	y32	y33	y34	y35
y36	y37	y38	y39	y40	y41	y42
y43	y44	y45	y46	y47	y48	y49

x1	x2	x3	x4	x5	x6	x7
x8						k-x8
x9						k-x9
x10						k-x10
x11						k-x11
x12						k-x12
k-x7	k-x2	k-x3	k-x4	k-x5	k-x6	k-x1

x13	x14	x15	x16	x17	x18	x19
x20						k-x20
x21						k-x21
x22						k-x22
x23						k-x23
x24						k-x24
k-x19	k-x14	k-x15	k-x16	k-x17	k-x18	k-x13

x25	x26	x27	x28	x29	x30	x31
x32						k-x32
x33						k-x33
x34			k/2			k-x34
x35						k-x35
x36						k-x36
k-x31	k-x26	k-x27	k-x28	k-x29	k-x30	k-x25

x37	x38	x39	x40	x41	x42	x43
x44						k-x44
x45						k-x45
x46						k-x46
x47						k-x47
x48						k-x48
k-x43	k-x38	k-x39	k-x40	k-x41	k-x42	k-x37

x49	x50	x51	x52	x53	x54	x55
x56						k-x56
x57						k-x57
x58						k-x58
x59						k-x59
x60						k-x60
k-x55	k-x50	k-x51	k-x52	k-x53	k-x54	k-x49

k-y49	k-y44	k-y45	k-y46	k-y47	k-y48	k-y43
k-y14	k-y9	k-y10	k-y11	k-y12	k-y13	k-8
k-y21	k-y16	k-y17	k-y18	k-y19	k-y20	k-y15
k-y28	k-y23	k-y24	k-y25	k-y26	k-y27	k-y22
k-y35	k-y30	k-y31	k-y32	k-y33	k-y34	k-y29
k-y42	k-y37	k-y38	k-y39	k-y40	k-y41	k-y36
k-y7	k-y2	k-y3	k-y4	k-y5	k-y6	k-y1

The magic constant is $S = 7k / 2$.

In general, the interior can be any magic cube of order 5 with magic constants $S = 5k / 2$, but it can also be a concentric magic cube.

Figure 22 contains an example of a classical concentric magic cube where the inside is not a concentric magic cube of order 5. This cube is made up of distinct positive integers from 1 to 343.

The interior of the cube is an unconventional associative pantriagonal cube of order 5 with a magic constant $S = 860$.

Figure 22

$S = 1204, k = 344$

318	8	57	261	7	227	326
312	11	60	258	10	222	331
17	305	163	103	136	324	156
51	137	170	169	299	196	182
91	292	245	267	161	84	64
105	235	306	50	300	66	142
310	216	203	96	291	85	3

200	73	111	123	255	176	266
1	296	230	199	133	2	343
304	28	322	256	160	94	40
72	120	54	283	217	186	272
219	212	81	15	309	243	125
330	204	173	107	41	335	14
78	271	233	221	89	168	144

116	97	79	25	320	254	313
340	223	192	126	30	289	4
110	315	249	153	87	56	234
274	82	276	210	179	113	70
215	74	43	302	236	205	129
118	166	100	69	328	197	226
31	247	265	319	24	90	228

19	242	187	59	323	180	194
214	185	119	23	282	251	130
195	277	146	80	49	308	149
193	269	238	172	106	75	151
232	36	295	264	198	67	112
201	93	62	321	225	159	143
150	102	157	285	21	164	325

33	339	297	268	131	71	65
135	147	16	275	244	178	209
298	139	108	42	301	270	46
206	231	165	134	68	262	138
63	288	257	191	95	29	281
190	55	314	218	152	121	154
279	5	47	76	213	273	311

177	317	332	220	115	37	6
189	9	303	237	171	140	155
92	101	35	329	263	132	252
246	158	127	61	290	224	98
104	250	184	88	22	316	240
58	342	211	145	114	48	286
338	27	12	124	229	307	167

341	128	141	248	53	259	34
13	333	284	86	334	122	32
188	39	181	241	208	20	327
162	207	174	175	45	148	293
280	52	99	77	183	260	253
202	109	38	294	44	278	239
18	336	287	83	337	117	26

Next, we will consider the case where a concentric magic cube of order 5 is inside a concentric magic cube of order 7.

The scheme of this magic cube is shown in Fig. 23.

Figure 23

y1	y2	y3	y4	y5	y6	y7
y8	y9	y10	y11	y12	y13	y14
y15	y16	y17	y18	y19	y20	y21
y22	y23	y24	y25	y26	y27	y28
y29	y30	y31	y32	y33	y34	y35
y36	y37	y38	y39	y40	y41	y42
y43	y44	y45	y46	y47	y48	y49

x1	x2	x3	x4	x5	x6	x7
x8	z1	z2	z3	z4	z5	k-x8
x9	z6	z7	z8	z9	z10	k-x9
x10	z11	z12	z13	z14	z15	k-x10
x11	z16	z17	z18	z19	z20	k-x11
x12	z21	z22	z23	z24	z25	k-x12
k-x7	k-x2	k-x3	k-x4	k-x5	k-x6	k-x1

x13	x14	x15	x16	x17	x18	x19
x20	z26	z27	z28	z29	z30	k-x20
x21	z31				k-z31	k-x21
x22	z32				k-z32	k-x22
x23	z33				k-z33	k-x23
x24	k-z30	k-z27	k-z28	k-z29	k-z26	k-x24
k-x19	k-x14	k-x15	k-x16	k-x17	k-x18	k-x13

x25	x26	x27	x28	x29	x30	x31
x32	z34	z35	z36	z37	z38	k-x32
x33	z39				k-z39	k-x33
x34	z40		k/2		k-z40	k-x34
x35	z41				k-z41	k-x35
x36	k-z38	k-z35	k-z36	k-z37	k-z34	k-x36
k-x31	k-x26	k-x27	k-x28	k-x29	k-x30	k-x25

x37	x38	x39	x40	x41	x42	x43
x44	z42	z43	z44	z45	z46	k-x44
x45	z47				k-z47	k-x45
x46	z48				k-z48	k-x46
x47	z49				k-z49	k-x47
x48	k-z46	k-z43	k-z44	k-z45	k-z42	k-x48
k-x43	k-x38	k-x39	k-x40	k-x41	k-x42	k-x37

x49	x50	x51	x52	x53	x54	x55
x56	k-z25	k-z22	k-z23	k-z24	k-z21	k-x56
x57	k-z10	k-z7	k-z8	k-z9	k-z6	k-x57
x58	k-z15	k-z12	k-z13	k-z14	k-z11	k-x58
x59	k-z20	k-z17	k-z18	k-z19	k-z16	k-x59
x60	k-z5	k-z2	k-z3	k-z4	k-z1	k-x60
k-x55	k-x50	k-x51	k-x52	k-x53	k-x54	k-x49

k-y49	k-y44	k-y45	k-y46	k-y47	k-y48	k-y43
k-y14	k-y9	k-y10	k-y11	k-y12	k-y13	k-8
k-y21	k-y16	k-y17	k-y18	k-y19	k-y20	k-y15
k-y28	k-y23	k-y24	k-y25	k-y26	k-y27	k-y22
k-y35	k-y30	k-y31	k-y32	k-y33	k-y34	k-y29
k-y42	k-y37	k-y38	k-y39	k-y40	k-y41	k-y36
k-y7	k-y2	k-y3	k-y4	k-y5	k-y6	k-y1

The magic constant of this concentric cube is $S = 7k / 2$.

Inside the concentric cube of order 7 is a concentric cube of order 5 with a magic constant

$S = 5k / 2$. Furthermore, inside the concentric cube of order 5 is a magic cube of order 3 with a magic constant $S = 3k / 2$.

Our first step is to show the classic concentric magic cube of order 7 of this type, and it appears in figure 24.

Figure 24

S = 1204, k = 344

17	206	28	315	41	325	272
58	295	139	87	231	194	200
191	328	143	228	79	137	98
190	147	326	232	77	135	97
189	133	141	234	323	75	109
238	51	275	70	174	221	175
321	44	152	38	279	117	253

212	73	94	101	309	125	290
131	26	102	230	229	273	213
168	119	259	254	104	124	176
261	262	118	103	255	122	83
249	140	145	146	151	278	95
129	313	236	127	121	63	215
54	271	250	243	35	219	132

99	84	78	46	299	276	322
301	40	3	142	337	338	43
287	148	182	333	1	196	57
42	332	310	20	186	12	302
159	334	24	163	329	10	185
294	6	341	202	7	304	50
22	260	266	298	45	68	245

156	76	244	105	297	165	161
285	173	308	107	62	210	59
86	216	319	2	195	128	258
284	30	48	172	296	314	60
74	307	149	342	25	37	270
136	134	36	237	282	171	208
183	268	100	239	47	179	188

312	160	280	251	67	53	81
52	340	339	164	9	8	292
130	157	15	181	320	187	214
92	14	158	324	34	330	252
178	13	343	11	162	331	166
177	336	5	180	335	4	167
263	184	64	93	277	291	32

317	305	288	80	126	33	55
233	281	108	217	223	31	111
96	220	85	90	240	225	248
88	222	226	241	89	82	256
120	66	199	198	193	204	224
61	71	242	114	115	318	283
289	39	56	264	218	311	27

91	300	192	306	65	227	23
144	49	205	257	113	150	286
246	16	201	116	265	207	153
247	197	18	112	267	209	154
235	211	203	110	21	269	155
169	293	69	274	170	123	106

A concentric magic cube of order 7 of distinct primes appears in figure 25.

This magic cube is not easy to make. Inside is a concentric magic cube of order 5 with a magic constant S = 54515. Inside the cube of order 5 is a magic cube of order 3 with a magic constant S = 32709.

Figure 25

S = 76321, k = 21806

1783	19429	2593	3373	14983	19423	14737
1399	13417	17203	17047	6883	19	20353
13267	14629	15727	859	4567	13339	13933
20347	3943	19249	1483	19777	4903	6619
787	11353	17383	17293	1033	10303	18169
20809	4507	1693	17713	11119	19267	1213
17929	9043	2473	18553	17959	9067	1297

9103	3583	5107	5737	18097	18757	15937
823	14347	21163	9349	4099	5557	20983
16993	3793	9883	20899	9463	10477	4813
15259	6379	15739	12919	10369	9109	6547
18307	13627	3673	3019	21397	12799	3499
9967	16369	4057	8329	9187	16573	11839
5869	18223	16699	16069	3709	3049	12703

6247	4363	4297	21319	20743	1993	17359
12073	4639	7	6607	21523	21739	9733
13879	7723	9769	21613	1327	14083	7927
8839	21727	20107	229	12373	79	12967
19753	20359	2833	10867	19009	1447	2053
11083	67	21799	15199	283	17167	10723
4447	17443	17509	487	1063	19813	15559

919	5743	14107	18199	13597	9319	14437
19717	8803	523	18919	12757	13513	2089
17209	13177	20143	157	12409	8629	4597
9973	13093	3169	10903	18637	8713	11833
16057	11149	9397	21649	1663	10657	5749
5077	8293	21283	2887	9049	13003	16729
7369	16063	7699	3607	8209	12487	20887

16693	17257	15607	13873	3967	8647	277
20929	21493	15073	6163	3517	8269	877
307	18493	2797	10939	18973	3313	21499
1987	619	9433	21577	1699	21187	19819
12157	373	20479	193	12037	21433	9649
2719	13537	6733	15643	18289	313	19087
21529	4549	6199	7933	17839	13159	5113

21067	13183	15277	10567	1087	5443	9697
19927	5233	17749	13477	12619	5437	1879
6793	11329	11923	907	12343	18013	15013
4729	12697	6067	8887	11437	15427	17077
5623	9007	18133	18787	409	8179	16183
6073	16249	643	12457	17707	7459	15733
12109	8623	6529	11239	20719	16363	739

20509	12763	19333	3253	3847	12739	3877
1453	8389	4603	4759	14923	21787	20407
7873	7177	6079	20947	17239	8467	8539
15187	17863	2557	20323	2029	16903	1459
3637	10453	4423	4513	20773	11503	21019
20593	17299	20113	4093	10687	2539	997
7069	2377	19213	18433	6823	2383	20023

This concentric magic cube is shown in the picture "Russian nesting dolls" (see Fig. 26).

Figure 26

 An attempt was made to make a concentric magic cube of order 7 of distinct primes with magic constant S < 76321. My attempts were not successful. I show two partial solutions (Fig. 27 and Fig. 28).

 A challenge for the readers - find the edging! There is no guarantee that the problem has a solution.

Figure 27

S = 69769, k = 19934

y1	y2	y3	y4	y5	y6	y7
y8	y9	y10	y11	y12	y13	y14
y15	y16	y17	y18	y19	y20	y21
y22	y23	y24	y25	y26	y27	y28
y29	y30	y31	y32	y33	y34	y35
y36	y37	y38	y39	y40	y41	y42
y43	y44	y45	y46	y47	y48	y49

x1	x2	x3	x4	x5	x6	x7
x8	2857	7237	17683	5641	16417	k-x8
x9	8761	4861	5827	15733	14653	k-x9
x10	7333	13063	15271	14011	157	k-x10
x11	15607	16747	4027	1483	11971	k-x11
x12	15277	7927	7027	12967	6637	k-x12
k-x7	k-x2	k-x3	k-x4	k-x5	k-x6	k-x1

x13	x14	x15	x16	x17	x18	x19
x20	2143	7	8431	19387	19867	k-x20
x21	10333	12541	17317	43	9601	k-x21
x22	18211	14767	4483	10651	1723	k-x22
x23	19081	2593	8101	19207	853	k-x23
x24	67	19927	11503	547	17791	k-x24
k-x19	k-x14	k-x15	k-x16	k-x17	k-x18	k-x13

x25	x26	x27	x28	x29	x30	x31
x32	11941	10831	8803	17377	883	k-x32
x33	14947	16633	751	12517	4987	k-x33
x34	73	5851	9967	14083	19861	k-x34
x35	3823	7417	19183	3301	16111	k-x35
x36	19051	9103	11131	2557	7993	k-x36
k-x31	k-x26	k-x27	k-x28	k-x29	k-x30	k-x25

x37	x38	x39	x40	x41	x42	x43
x44	19597	19753	2011	463	8011	k-x44
x45	10513	727	11833	17341	9421	k-x45
x46	4441	9283	15451	5167	15493	k-x46
x47	3361	19891	2617	7393	16573	k-x47
x48	11923	181	17923	19471	337	k-x48
k-x43	k-x38	k-x39	k-x40	k-x41	k-x42	k-x37

x49	x50	x51	x52	x53	x54	x55
x56	13297	12007	12907	6967	4657	k-x56
x57	5281	15073	14107	4201	11173	k-x57
x58	19777	6871	4663	5923	12601	k-x58
x59	7963	3187	15907	18451	4327	k-x59
x60	3517	12697	2251	14293	17077	k-x60
k-x55	k-x50	k-x51	k-x52	k-x53	k-x54	k-x49

k-y49	k-y44	k-y45	k-y46	k-y47	k-y48	k-y43
k-y14	k-y9	k-y10	k-y11	k-y12	k-y13	k-8
k-y21	k-y16	k-y17	k-y18	k-y19	k-y20	k-y15
k-y28	k-y23	k-y24	k-y25	k-y26	k-y27	k-y22
k-y35	k-y30	k-y31	k-y32	k-y33	k-y34	k-y29
k-y42	k-y37	k-y38	k-y39	k-y40	k-y41	k-y36
k-y7	k-y2	k-y3	k-y4	k-y5	k-y6	k-y1

Figure 28

S = 68999, k = 19714

y1	y2	y3	y4	y5	y6	y7
y8	y9	y10	y11	y12	y13	y14
y15	y16	y17	y18	y19	y20	y21
y22	y23	y24	y25	y26	y27	y28
y29	y30	y31	y32	y33	y34	y35
y36	y37	y38	y39	y40	y41	y42
y43	y44	y45	y46	y47	y48	y49

x1	x2	x3	x4	x5	x6	x7
x8	16253	2141	14747	8501	7643	k-x8
x9	2087	18617	14771	8093	5717	k-x9
x10	1877	11831	4073	19463	12041	k-x10
x11	17891	3083	2243	12671	13397	k-x11
x12	11177	13613	13451	557	10487	k-x12
k-x7	k-x2	k-x3	k-x4	k-x5	k-x6	k-x1

x13	x14	x15	x16	x17	x18	x19
x20	1667	1637	7001	19403	19577	k-x20
x21	8627	7817	18731	3023	11087	k-x21
x22	19421	17417	1373	10781	293	k-x22
x23	19433	4337	9467	15767	281	k-x23
x24	137	18077	12713	311	18047	k-x24
k-x19	k-x14	k-x15	k-x16	k-x17	k-x18	k-x13

x25	x26	x27	x28	x29	x30	x31
x32	2927	19697	12911	1523	12227	k-x32
x33	16871	17807	593	11171	2843	k-x33
x34	17783	3221	9857	16493	1931	k-x34
x35	4217	8543	19121	1907	15497	k-x35
x36	7487	17	6803	18191	16787	k-x36
k-x31	k-x26	k-x27	k-x28	k-x29	k-x30	k-x25

x37	x38	x39	x40	x41	x42	x43
x44	19211	19709	8363	701	1301	k-x44
x45	7703	3947	10247	15377	12011	k-x45
x46	2531	8933	18341	2297	17183	k-x46
x47	1427	16691	983	11897	18287	k-x47
x48	18413	5	11351	19013	503	k-x48
k-x43	k-x38	k-x39	k-x40	k-x41	k-x42	k-x37

x49	x50	x51	x52	x53	x54	x55
x56	9227	6101	6263	19157	8537	k-x56
x57	13997	1097	4943	11621	17627	k-x57
x58	7673	7883	15641	251	17837	k-x58
x59	6317	16631	17471	7043	1823	k-x59
x60	12071	17573	4967	11213	3461	k-x60
k-x55	k-x50	k-x51	k-x52	k-x53	k-x54	k-x49

k-y49	k-y44	k-y45	k-y46	k-y47	k-y48	k-y43
k-y14	k-y9	k-y10	k-y11	k-y12	k-y13	k-8
k-y21	k-y16	k-y17	k-y18	k-y19	k-y20	k-y15
k-y28	k-y23	k-y24	k-y25	k-y26	k-y27	k-y22
k-y35	k-y30	k-y31	k-y32	k-y33	k-y34	k-y29
k-y42	k-y37	k-y38	k-y39	k-y40	k-y41	k-y36
k-y7	k-y2	k-y3	k-y4	k-y5	k-y6	k-y1

Concentric magic cubes of order 8

On the web page

http://www.magic-SquareS.net/c-t-htm/c_prime.htm

you will find the concentric magic cube of order 8 of distinct primes that appears in figure 29.

The magic constant of the cube is S = 39480. Inside is a concentric magic cube of order 6 with a magic constant S = 29610, this is the magic cube that appeared in figure 1. Inside the concentric cube of order 6 is an associative and pantriagonal cube of order 4 with a magic constant

S = 19740.

Figure 29

S = 39480, k = 9870

13	9859	6679	9829	2129	53	6869	4049
1637	9781	103	8171	181	7577	9733	2297
9511	349	3623	269	433	9787	7691	7817
9631	257	7331	2477	9371	9413	521	479
9283	1039	941	631	8837	661	8861	9227
6803	709	3613	8443	9187	3541	2617	4567
1493	8707	9043	907	8291	6701	1171	3167
1109	8779	8147	8753	1051	1747	2017	7877

811	9127	7841	5867	7211	2909	3931	1783
6781	4831	4783	67	9811	4639	5479	3089
4229	191	241	193	9473	9769	9743	5641
409	331	577	5009	4751	9619	9323	9461
7177	8273	9719	8933	1123	829	733	2693
4967	8423	7499	8287	1789	1801	1811	4903
7019	7561	6791	7121	2663	2953	2521	2851
8087	743	2029	4003	2659	6961	5939	9059

8431	1289	4951	4933	1063	8941	9013	859
5717	131	761	379	9403	9497	9439	4153
6151	8951	2437	3547	5309	8447	919	3719
19	9643	3209	5573	2281	8677	227	9851
2711	2143	8243	4877	6007	613	7727	7159
3307	8311	5851	5743	6143	2003	1559	6563
4133	431	9109	9491	467	373	9739	5737
9011	8581	4919	4937	8807	929	857	1439

8783	1181	8093	1759	1933	6379	2633	8719
4201	337	8849	8821	1409	1307	8887	5669
3671	7013	5903	2879	9007	1951	2857	6199
8641	8009	3217	2767	8117	5639	1861	1229
7523	9049	6073	5521	2333	5813	821	2347
2719	4219	4547	8573	283	6337	5651	7151
2791	983	1021	1049	8461	8563	9533	7079
1151	8689	1777	8111	7937	3491	7237	1087

8669	1223	1483	7583	2267	7477	5197	5581
3779	8543	8839	9277	173	1831	947	6091
3917	4177	3533	9587	1297	5323	5693	5953
5881	7487	4057	7537	4349	3797	2383	3989
2381	31	4231	1753	7103	6653	9839	7489
3803	449	7919	863	6991	3967	9421	6067
6761	8923	1031	593	9697	8039	1327	3109
4289	8647	8387	2287	7603	2393	4673	1201

7529	7993	2111	3041	7789	3889	3947	3181
3449	8419	3299	8317	1607	5419	2549	6421
4021	9151	7867	3727	4127	4019	719	5849
1481	3593	9257	3863	4993	1627	6277	8389
2113	977	1193	7589	4297	6661	8893	7757
7129	149	1423	4561	6323	7433	9721	2741
7069	7321	6571	1553	8263	4451	1451	2801
6689	1877	7759	6829	2081	5981	5923	2341

3251	7717	6599	5351	8269	1709	37	6547
6343	7349	3079	2749	7207	6917	2309	3527
5927	127	9629	9677	397	101	9679	3943
4027	547	9293	4861	5119	251	9539	5843
7649	9137	151	937	8747	9041	1597	2221
5449	8059	2371	1583	8081	8069	1447	4421
3511	4391	5087	9803	59	5231	5039	6359
3323	2153	3271	4519	1601	8161	9833	6619

1993	1091	1723	1117	8819	8123	7853	8761
7573	89	9767	1699	9689	2293	137	8233
2053	9521	6247	9601	9437	83	2179	359
9391	9613	2539	7393	499	457	9349	239
643	8831	8929	9239	1033	9209	1009	587
5303	9161	6257	1427	683	6329	7253	3067
6703	1163	827	8963	1579	3169	8699	8377
5821	11	3191	41	7741	9817	3001	9857

The scheme for the concentric magic cube of order 8, based on the standard cube depicted in Fig. 29, appears in figure 30.

Figure 30

S = 4k

x1	x2	x3	x4	x5	x6	x7	x8
x9	x10	x11	x12	x13	x14	x15	x16
x17	x18	x19	x20	x21	x22	x23	x24
x25	x26	x27	x28	x29	x30	x31	x32
x33	x34	x35	x36	x37	x38	x39	x40
x41	x42	x43	x44	x45	x46	x47	x48
x49	x50	x51	x52	x53	x54	x55	x56
x57	x58	x59	x60	x61	x62	x63	x64

x65	x66	x67	x68	x69	x70	x71	x72
x73	z1	z2	z3	z4	z5	z6	k-x73
x74	z7	z8	z9	z10	z11	z12	k-x74
x75	z13	z14	z15	z16	z17	z18	k-x75
x76	z19	z20	z21	z22	z23	z24	k-x76
x77	z25	z26	z27	z28	z29	z30	k-x77
x78	z31	z32	z33	z34	z35	z36	k-x78
k-x72	k-x66	k-x67	k-x68	k-x69	k-x70	k-x71	k-x65

x79	x80	x81	x82	x83	x84	x85	x86
x87	z37	z38	z39	z40	z41	z42	k-x87
x88	z43					k-z43	k-x88
x89	z44					k-z44	k-x89
x90	z45					k-z45	k-x90
x91	z46					k-z46	k-x91
x92	k-z42	k-z38	k-z39	k-z40	k-z41	k-z37	k-x92
k-x86	k-x80	k-x81	k-x82	k-x83	k-x84	k-x85	k-x79

x93	x94	x95	x96	x97	x98	x99	x100
x101	z47	z48	z49	z50	z51	z52	k-x101
x102	z53					k-z53	k-x102
x103	z54					k-z54	k-x103
x104	z55					k-z55	k-x104
x105	z56					k-z56	k-x105
x106	k-z52	k-z48	k-z49	k-z50	k-z51	k-z47	k-x106
k-x100	k-x94	k-x95	k-x96	k-x97	k-x98	k-x99	k-x93

x107	x108	x109	x110	x111	x112	x113	x114
x115	z57	z58	z59	z60	z61	z62	k-x115
x116	z63					k-z63	k-x116
x117	z64					k-z64	k-x117
x118	z65					k-z65	k-x118
x119	z66					k-z66	k-x119
x120	k-z62	k-z58	k-z59	k-z60	k-z61	k-z57	k-x120
k-x114	k-x108	k-x109	k-x110	k-x111	k-x112	k-x113	k-x107

x121	x122	x123	x124	x125	x126	x127	x128
x129	z67	z68	z69	z70	z71	z72	k-x129
x130	z73					k-z73	k-x130
x131	z74					k-z74	k-x131
x132	z75					k-z75	k-x132
x133	z76					k-z76	k-x133
x134	k-z72	k-z68	k-z69	k-z70	k-z71	k-z67	k-x134
k-x128	k-x122	k-x123	k-x124	k-x125	k-x126	k-x127	k-x121

x135	x136	x137	x138	x139	x140	x141	x142
x143	k-z36	k-z32	k-z33	k-z34	k-z35	k-z31	k-x143
x144	k-z12	k-z8	k-z9	k-z10	k-z11	k-z7	k-x144
x145	k-z18	k-z14	k-z15	k-z16	k-z17	k-z13	k-x145
x146	k-z24	k-z20	k-z21	k-z22	k-z23	k-z19	k-x146
x147	k-z30	k-z26	k-z27	k-z28	k-z29	k-z25	k-x147
x148	k-z6	k-z2	k-z3	k-z4	k-z5	k-z1	k-x148
k-x142	k-x136	k-x137	k-x138	k-x139	k-x140	k-x141	k-x135

k-x64	k-x58	k-x59	k-x60	k-x61	k-x62	k-x63	k-x57
k-x16	k-x10	k-x11	k-x12	k-x13	k-x14	k-x15	k-x9
k-x24	k-x18	k-x19	k-x20	k-x21	k-x22	k-x23	k-x17
k-x32	k-x26	k-x27	k-x28	k-x29	k-x30	k-x31	k-x25
k-x40	k-x34	k-x35	k-x36	k-x37	k-x38	k-x39	k-x33
k-x48	k-x42	k-x43	k-x44	k-x45	k-x46	k-x47	k-x41
k-x56	k-x50	k-x51	k-x52	k-x53	k-x54	k-x55	k-x49
k-x8	k-x2	k-x3	k-x4	k-x5	k-x6	k-x7	k-x1

An attempt was made to make this scheme a concentric magic cube of order 8 of distinct primes. The best result was the incomplete solution that appears in figure 31.

76

Figure 31

S = 33600, k = 8400

4817	2819	2269	7517	3067	4871	4783	3457
4889	2741	2137	7523	2963	4909	4729	3709
6067	2557	5009	5281	1373	7541	839	4933
7019	1619	7349	1249	4289	1931	7247	2897
5849	3319	8093	3833	5743	3253	1709	1801
1951	6829	647	4273	3919	2617	5647	7717
2731	5483	5557	3167	5039	797	6949	3877
277	8233	2539	757	7207	7681	1697	5209

x65	x66	x67	x68	x69	x70	x71	x72
x73	5197	5521	5171	4373	677	4261	k-x73
x74	2311	2699	7237	1361	6029	5563	k-x74
x75	6803	919	4787	8053	311	4327	k-x75
x76	5087	7937	1543	1747	7433	1453	k-x76
x77	571	2267	3413	3469	8171	7309	k-x77
x78	5231	5857	3049	6197	2579	2287	k-x78
k-x72	k-x66	k-x67	k-x68	k-x69	k-x70	k-x71	k-x65

x79	x80	x81	x82	x83	x84	x85	x86
x87	523	809	11	8269	8011	7577	k-x87
x88	37	1213	8293	2621	4673	8363	k-x88
x89	8161	7573	643	5807	2777	239	k-x89
x90	8287	6911	137	2011	7741	113	k-x90
x91	7369	1103	7727	6361	1609	1031	k-x91
x92	823	7591	8389	131	389	7877	k-x92
k-x86	k-x80	k-x81	k-x82	k-x83	k-x84	k-x85	k-x79

x93	x94	x95	x96	x97	x98	x99	x100
x101	179	6763	5717	3389	5689	3463	k-x101
x102	8317	709	6397	6287	3407	83	k-x102
x103	2503	8387	1399	2153	4861	5897	k-x103
x104	2143	727	7823	31	8219	6257	k-x104
x105	7121	6977	1181	8329	313	1279	k-x105
x106	4937	1637	2683	5011	2711	8221	k-x106
k-x100	k-x94	k-x95	k-x96	k-x97	k-x98	k-x99	k-x93

x107	x108	x109	x110	x111	x112	x113	x114
x115	4951	7537	1667	1567	4241	5237	k-x115
x116	7901	8087	71	7219	1423	499	k-x116
x117	1187	181	8369	577	7673	7213	k-x117
x118	1297	3539	6247	7001	13	7103	k-x118
x119	6701	4993	2113	2003	7691	1699	k-x119
x120	3163	863	6733	6833	4159	3449	k-x120
k-x114	k-x108	k-x109	k-x110	k-x111	k-x112	k-x113	k-x107

x121	x122	x123	x124	x125	x126	x127	x128
x129	8237	2027	7283	5399	761	1493	k-x129
x130	3797	6791	2039	673	7297	4603	k-x130
x131	2473	659	6389	8263	1489	5927	k-x131
x132	1439	5623	2593	7757	827	6961	k-x132
x133	2347	3727	5779	107	7187	6053	k-x133
x134	6907	6373	1117	3001	7639	163	k-x134
k-x128	k-x122	k-x123	k-x124	k-x125	k-x126	k-x127	k-x121

x135	x136	x137	x138	x139	x140	x141	x142
x143	6113	2543	5351	2203	5821	3169	k-x143
x144	2837	5701	1163	7039	2371	6089	k-x144
x145	4073	7481	3613	347	8089	1597	k-x145
x146	6947	463	6857	6653	967	3313	k-x146
x147	1091	6133	4987	4931	229	7829	k-x147
x148	4139	2879	3229	4027	7723	3203	k-x148
k-x142	k-x136	k-x137	k-x138	k-x139	k-x140	k-x141	k-x135

3191	167	5861	7643	1193	719	6703	8123
4691	5659	6263	877	5437	3491	3671	3511
3467	5843	3391	3119	7027	859	7561	2333
5503	6781	1051	7151	4111	6469	1153	1381
6599	5081	307	4567	2657	5147	6691	2551
683	1571	7753	4127	4481	5783	2753	6449
4523	2917	2843	5233	3361	7603	1451	5669

A challenge is issued to the reader, find a complete solution.

Dear readers!
Should you find solutions to the proposed problems, please send them to the site

http://primesmagicgames.altervista.org/wp/competitions/

Furthermore, if you find a better solution, please send it in.

Links

1. http://en.wikipedia.org/wiki/Magic_cube

2. http://www.magic-SquareS.net/c-t-htm/c_prime.htm

3. http://primesmagicgames.altervista.org/wp/

4. Full size picture Fig. 26:

http://s017.radikal.ru/i427/1410/01/9dac57fb20e9.jpg

CONCATENATION PROBLEMS

Henry Ibstedt

Glimminge 2036

28060 Broby

Sweden

henry.ibstedt@gmail.com

Abstract

This study has been inspired by questions asked by Charles Ashbacher in the *Journal of Recreational Mathematics, vol. 29.2* concerning the Smarandache Deconstructive Sequence. This sequence is a special case of a more general concatenation and sequencing procedure which is the subject of this study. The properties of this kind of sequences are studied with particular emphasis on the divisibility of their terms by primes.

Introduction

In this article the concatenation of a and b is expressed by a_b or simply ab where there can be no misunderstanding. Multiple concatenations like abcabcabc will be expressed by 3(abc).

We consider n different elements (or n objects) arranged (concatenated) one after the other in the following way to form:

$A = a_1 a_2 \ldots a_n$.

Infinitely many objects A, which will be referred to as cycles, are concatenated to form the chain:

$B = a_1 a_2 \ldots a_n \ a_1 a_2 \ldots a_n \ a_1 a_2 \ldots a_n \ldots$

B contains identical elements which are at equidistant positions in the chain. Let's write B as

$B = b_1 b_2 b_3, \ldots b_k \ldots$ where $b_k = a_j$ when $j \equiv k \pmod{n}$, $1 \leq j \leq n$.

An infinite sequence $C_1, C_2, C_3, \ldots C_k, \ldots$ is formed by sequentially selecting 1, 2, 3, …k, … elements from the chain B:

$C_1 = b_1 = a_1$

$C_2 = b_2 b_3 = a_2 a_3$

$C_3 = b_4 b_5 b_6 = a_4 a_5 a_6$ (if $n \leq 6$, if $n=5$ we would have $C_3 = a_4 a_5 a_1$)

The number of elements from the chain B used to form the first k-1 terms of the sequence C is $1+2+3+ \ldots +k-1 = (k-1)k/2$. Hence

$$C_k = b_{\frac{(k-1)k}{2}+1} \, b_{\frac{(k-1)k}{2}+2} \, \cdots \, b_{\frac{k(k+1)}{2}}$$

However, what is interesting to see is how C_k is expressed in terms of $a_1, \ldots, a_n$. For sufficiently large values of k, C_k will be composed of three parts:

The first part $F(k) = a_u \ldots a_n$

The middle part $M(k) = AA \ldots A$. The number of concatenated A's depends on k.

The last part $L(k) = a_1 a_2 \ldots a_w$

Hence

(1) $C_k = F(k) M(k) L(k)$.

The number of elements used to form $C_1, C_2, \ldots C_{k-1}$ is $((k-1)k)/2$. Since the number of elements in A is finite there will be infinitely many terms C_k which have the same first element a_u. u can be determined from

$$\frac{(k-1)k}{2} + 1 \equiv u(\bmod n)$$

There can be at most n^2 different combinations to form F(k) and L(k).

Let C_j and C_i be two different terms for which F(i)=F(j) and L(i)=L(j). They will then be separated by a number m of complete cycles of length n, i.e.

$$\frac{(j-1)j}{2} - \frac{(i-1)i}{2} = mn$$

Let's write j=i+p and see if p exists so that there is a solution for p which is independent of i.

$$(i+p-1)(i+p)-(i-1)i=2mn$$

$$p^2+p(2i-1)=2mn$$

If n is odd we will put p=n to obtain n+2i-1=2m, or

$$m = \frac{n+2i-1}{2}$$

If n is even we put p=2n to obtain m=2n+2i-1. From this we see that the terms C_k have a peculiar periodic behaviour. The periodicity is p=n for odd n and p=2n for even n. Let's illustrate this for n=4 and n=5 for which the periodicity will be p=8 and p=5 respectively. It is seen from table 1 that the periodicity starts for i=3.

Numerals are chosen as elements to illustrate the case n=5 (table 2). Let's write i=s+k+pj , where s is the index of the term preceding the first periodical term, k=1,2,...,p is the index of members of the period and j is the number of the period (for convenience the first period is numbered 0). The first part of C_i is denoted B(k) and the last part E(k). C_i is now given by the following expression, where q is the number of cycles concatenated between the first part B(k) and the last part E(k).

(2) C_i=B(k)_qA_E(k), where k is determined from i-s≡k (mod p)

84

Table 1

n=4. A=abcd. B= abcdabcdabcdabcdabcd……

	C_I	Period #	F(i)	M(i)	L(i)
1	a		a		
2	bc		bc		
3	dab	1	d		ab
4	cdab	1	cd		ab
5	cdabc	1	cd		abc
6	dabcda	1	d	abcd	a
7	bcdabcd	1	bcd	abcd	
8	abcdabcd	1		2(abcd)	
9	abcdabcda	1		2(abcd)	a
10	bcdabcdabc	1	bcd	abcd	abc
11	dabcdabcdab	2	d	2(abcd)	ab
12	cdabcdabcdab	2	cd	2(abcd)	ab
13	cdabcdabcdabc	2	cd	2(abcd)	abc
14	dabcdabcdabcda	2	d	3(abcd)	a
15	bcdabcdabcdabcd	2	bcd	3(abcd)	
16	abcdabcdabcdabcd	2		4(abcd)	
17	abcdabcdabcdabcda	2		4(abcd)	a
18	bcdabcdabcdabcdabc	2	bcd	3(abcd)	abc
19	dabcdabcdabcdabcdab	3	d	4(abcd)	ab
20	cdabcdabcdabcdabcdab	3	cd	4(abcd)	ab

85

Table 2

n=5. A=12345. B= 123451234512345………

I	C_I	k	q	F(i)/B(k)	M(I)	L(i)/E(k)
1	1			1		
s=2	23			23		
	j=0					
3	451	1	0	45		1
4	2345	2	0	2345		
5	12345	3	1		12345	
6	123451	4	1		12345	1
7	2345123	5	0	2345		123
	j=1					
3+5j	45123451	1	j	45	12345	1
4+5j	234512345	2	j	2345	12345	
5+5j	1234512345	3	j+1		2(12345)	
6+5j	12345123451	4	j+1		2(12345)	1
7+5j	234512345123	5	j	2345	12345	123
	j=2					
3+5j	4512345123451	1	j	45	2(12345)	1
4+5j	23451234512345	2	j	2345	2(12345)	
…						

86

The Smarandache Deconstructive Sequence

The Smarandache Deconstructive Sequence of integers [1] is constructed by sequentially repeating the digits 1-9 in the following way:

$$1,23,456,789123,4567891,23456789,123456789,1234567891, \ldots$$

The sequence was studied in a booklet by Kashihara [2] and a number of questions on this sequence were posed by Ashbacher [3]. In thinking about these questions two observations led to this study.

1. Why did Smarandache exclude 0 from the integers used to create the sequence? After all 0 is indispensible in all arithmetics most of which can be done using 0 and 1 only.

2. The process used to create the Deconstructive Sequence is a process which applies to any set of objects as has been shown in the introduction.

The periodicity and the general expression for terms in the "generalized deconstructive sequence" shown in the introduction may be the most important results of this study. These results will now be used to examine the questions raised by Ashbacher. It is worth noting that these divisibility questions are dealt with in base 10 although only the nine digits 1,2,3,4,5,6,7,8,9 are used to express numbers. In the last part of this article questions on divisibility will be posed for a deconstructive sequence generated from A="0123456789".

For $i > 5$ ($s = 5$) any term C_i in the sequence is composed by concatenating a first part $B(k)$, a number q of cycles A="123456789" and a last part $E(k)$, where $i = 5 + k + 9j$, k = 1,2,…9, $j \geq 0$, as expressed in (2) and q = j or j + 1 as shown in table 3.

Members of the Smarandache Deconstructive Sequence are now interpreted as decimal integers. The factorization of $B(k)$ and $E(k)$ is shown in table 3. The last two columns of this table will be useful later in this article.

Together with the factorization of the cycle $A=123456789=3^2 *3607 * 3803$ it is now possible to study some divisibility properties of the sequence. We will first find expressions for C_i for each of the 9 values of k. In cases where $E(k)$ exists let's introduce $u = 1 + [\log_{10}E(k)]$. We also define the function $\delta(j)$ so that $\delta(j) = 0$ for j = 0 and $\delta(j) = 1$ for j > 0. It is possible to construct one algorithm to cover all the nine cases but more functions like $\delta(j)$ would have to be introduced to distinguish between the numerical values of the strings "" (empty string) and "0" which are both evaluated as 0 in computer applications. In order to avoid this four formulas are used:

Table 3

Factorization of Smarandache Deconstructive Sequence

i	k	B(k)	q	E(k)	Digit sum	$3\|C_i$?
6+9j	1	789=3·263	j	123=3·41	30+j·45	3
7+9j	2	456789=3·43·3541	j	1	40+j·45	No
8+9j	3	23456789	j		44+j·45	No
9+9j	4		j+1		(j+1)·45	$9 \cdot 3^z$ *
10+9j	5		j+1	1	1+(j+1)·45	No
11+9j	6	23456789	j	123=341	50+j·45	No
12+9j	7	456789=3·43·3541	j	123456=2^6·3·643	60+j·45	3
13+9j	8	789=3·263	j+1	1	25+(j+1)·45	No
14+9j	9	23456789	j	123456=2^6·3·643	65+j·45	No

*) where z depends on j.

For k=1, 2, 6, 7 and 9:

$$(3) \quad C_{5+k+9j} = E(k) + \delta(j) * A * \sum_{r=0}^{j-1} 10^{9r} + B(k) * 10^{9j+u}.$$

For k=3:

$$(4) \quad C_{5+k+9j} = \delta(j) * A * \sum_{r=0}^{j-1} 10^{9r} + B(k)* 10^{9j}.$$

For k=4:

$$(5) \quad C_{5+k+9j} = A * \sum_{r=0}^{j} 10^{9r} .$$

For k=5 and 8:

$$(6) \quad C_{5+k+9j} = E(k) + A * 10^u * \sum_{r=0}^{j} 10^{9r} + B(k) * 10^{9(j+1)+u} .$$

Before dealing with the questions posed by Ashbacher we recall the familiar rules: An even number is divisible by 2; a number whose last two digit form a number which is divisible by 4 is divisible by 4.

In general we have the following:

Theorem. Let N be an n-digit integer such that $N > 2^{\alpha}$, then N is divisible by 2^{α} if and only if the number formed by the α last digits of N is divisible by 2^{α}.

Proof. To begin with we note that:

If x divides a and x divides b then x divides (a+b).

If x divides one but not the other of a and b then x does not divide (a+b).

If x does not divides neither a nor b then x may or may not divide (a+b).

Let's write the n-digit number in the form $a * 10^{\alpha} + b$. We then see from the following that $a * 10^{\alpha}$ is divisible by 2^{α}.

$$10 \equiv 0 \pmod{2}$$

$$100 \equiv 0 \pmod{4}$$

$$1000 = 2^3 * 5^3 \equiv 0 \pmod{2^3}$$

$$\dots$$

$$10^{\alpha} \equiv 0 \pmod{2^{\alpha}}$$

and then

$$a \cdot 10^{\alpha} \equiv 0 \pmod{2^{\alpha}} \text{ independent of a.}$$

Now let b be the number formed by the α last digits of N. We see from the introductory remark that N is divisible by 2^α if and only if the number formed by the α last digits is divisibele by 2^α.

Question 1. Does every even element of the Smarandache Deconstructive Sequence contain at least three instances of the prime 2 as a factor?

Question 2. If we form a sequence from the elements of the Smarandache Deconstructive Sequence that end in a 6, do the powers of 2 that divide them form a montonically increasing sequence?

These two questions are related and are dealt with together. From the previous analysis we know that all even elements of the Smarandache Deconstructive Sequence end in a 6. For $i \leq 5$ they are:

$$C_3 = 456 = 57 * 2^3$$

$$C_5 = 23456 = 733*2^5$$

For $i > 5$ they are of the forms:

$$C_{12+9j} \text{ and } C_{14+9j} \text{ which both end in } \ldots 789123456.$$

Examining the numbers formed by the 6, 7 and 8 last digits for divisibility by 2^6, 2^7 and 2^8 respectively we have:

$$123456 = 2^6 * 3 * 643$$

$$9123456 = 2^7 * 149 * 4673$$

$$89123456 \text{ is not divisible by } 2^8.$$

From this we conclude that all even Smarandache Deconstructive Sequence elements for $i \geq 12$ are divisible by 2^7 and that no elements in the sequence are divisible by higher powers of 2 than 7.

Answer to question 1. Yes.

Answer to question 2. The sequence is monotonically increasing for $i \leq 12$. For $i \geq 12$ the powers of 2 that divide even elements remain constant$=2^7$.

Question 3. Let x be the largest integer such that $3^x \mid i$ and y the largest integer such that $3^y \mid C_i$. Is it true that x is always equal to y?

From table 3 we se that the only elements C_i of the Smarandache Deconstructive Sequence which are divisible by powers of 3 correspond to i = 6 + 9j, 9 + 9j, or 12 + 9j. Furthermore, we see that $i = 6 + 9j$ and C_{6+9j} are divisible by 3 no more no less. The same is true for $i = 12 + 9j$ and C_{12+9j}. So the statement holds in these cases.

From the conguences

$$9 + 9j \equiv 0 \ (\text{mod } 3^x) \text{ for the index of the element}$$

and

$$45(1 + j) \equiv 0 \ (\text{mod } 3^y) \text{ for the corresponding element.}$$

we conclude that $x = y$.

Answer to question 3: The statement is true. It is interesting to note that, for example the 729 digit number C_{729} is divisible by 729.

Question 4. Are there other patterns of divisibility in this sequence?

A search for other patterns would continue by examining divisibility by the next lower primes 5, 7, 11, … It is obvious from table 3 and the periodicity of the sequence that there are no elements divisible by 5. The algorithms will prove very useful. For each value of k the value of C_i depends on j only. The divisibilty by a prime p is therefore determined by finding out for which values of j and k the congruence $C_i \equiv 0 \ (\text{mod } p)$ holds. We evaluate

$$\sum_{r=0}^{j-1} 10^{9r} = \frac{10^{9j} - 1}{10^9 - 1}.$$

and introduce $G = 10^9 - 1$. We note that $G = 3^4 * 37 * 333667$. From formulas (3) to (6) we now obtain:

For k=1,2,6,7 and 9:

(3′) $C_i * G = 10^u * (\delta(j) * A + B(k) * G) * 10^{9j} + E(k) * G - 10^u * \delta(j) * A.$

For k=3:

(4′) $C_i * G = ((\delta(j) * A + B(k) * G) * 10^{9j} - \delta(j) * A.$

For k=4:

(5′) $C_i * G = A * 10^{9j} - A.$

For k=5 and 8:

(6′) $C_i * G = 10^{u+9}(A + B(k) * G) * 10^{9j} + E(k) * G - 10^u * A.$

91

The divisibility of C_i by a prime p other than 3, 37 and 333667 is therefore determined by solutions for j to the congruences $C_iG \equiv 0 \pmod{p}$ which are of the form

(7) $a * (10^9)^j + b \equiv 0 \pmod{p}$.

Table 4 shows the results from computer implementation of the congruences. The appearance of elements divisible by a prime p is periodic, the periodicity is given by $j = j_1 + m * d$, $m = 1, 2, 3,$ ... . The first element divisible by p appears for i_1 corresponding to j_1. In general the terms C_i divisible by p are $C_{5+k+9(j1 + md)}$, where d is specific to the prime p and m=1, 2, 3,... .We note from table 4 that d is either equal to p - 1 or a divisor of p - 1 except for the case p=37 which as we have noted is a factor of A. Indeed this periodicity follows from Euler's extension of Fermat's little theorem because if we write (mod p):

$$a * (10^9)^j + b = a * (10^9)^{j_1+md} + b \equiv a * (10^9)^{j_1} + b \text{ for } d = p - 1 \text{ or a divisor of } p - 1.$$

Finally we note that the periodicity for p=37 is d=37.

Question: Table 4 indicates some interesting patterns. For instance, the primes 19, 43 and 53 only divide elements corresponding to k=1, 4 or 7 for j < 150 which was set as an upper limit for this study. Similarly, the primes 7,11, 41, 73, 79 and 91 only divides elements corresponding to k = 4. Is 5 the only prime that cannot divide an element of the Smarandache Deconstructive Sequence?

Table 4

Smarandache Deconstructive Sequence elements divisible by p

p=7	k	4
d=2	i_1	18
	j_1	1

p=11	k	4
d=2	i_1	18
	j_1	1

p=13	k	4	8	9
d=2	i_1	18	22	14
	j_1	1	1	0

p=17	k	1	2	3	4	5	6	7	8	9
d=16	i_1	6	43	44	144	100	101	138	49	95
	j_1	0	4	4	15	10	10	14	4	9

p=19	k	1	4	7
d=2	i_1	15	18	21
	j_1	1	1	1

p=23	k	1	2	3	4	5	6	7	8	9
d=22	i_1	186	196	80	198	118	200	12	184	14
	j_1	20	21	8	21	12	21	0	19	0

p=29	k	1	2	3	4	5	6	7	8	9
d=28	i_1	24	115	197	252	55	137	228	139	113
	j_1	2	12	21	27	5	14	24	14	11

p=31	k	3	4	5
d=5	i_1	26	45	19
	j_1	2	4	1

p=37	k	1	2	3	4	5	6	7	8	9
d=37	i_1	222	124	98	333	235	209	111	13	320
	j_1	24	13	10	36	25	22	11	0	34

p=41 d=5	k	4
	i_1	45
	j_1	4

P=43 d=7	1	4	7
	33	63	30
	3	6	2

p=47 d=46	k	1	2	3	4	5	6	7	8	9
	i_1	150	250	368	414	46	164	264	400	14
	j_1	16	27	40	45	4	17	28	43	0

p=53 d=13	k	1	4	7
	i_1	24	117	12
	j_1	2	12	9

p=59 d=58	k	1	3	5	6	7	8	9
	i_1	267	413	109	11	255	256	266
	j_1	29	45	11	0	27	27	28

p=61 d=20	k	2	4	6
	i_1	79	180	101
	j_1	8	19	10

p=67 d=11	k	4	8	9
	i_1	99	67	32
	j_1	10	6	2

p=71	k	1	3	4	5	7
d=35	i_1	114	53	315	262	201
	j_1	12	5	34	28	21

p=73	k	4
d=8	i_1	72
	j_1	7

p=79	k	4
d=13	i_1	117
	j_1	12

p=83	k	1	2	4	6	7	8	9
d=41	i_1	348	133	369	236	21	112	257
	j_1	38	14	40	25	1	11	27

p=89	k	2	4	6
d=44	i_1	97	396	299
	j_1	10	43	32

p=97	k	1	2	3	4	5	6	7	8	9
d=32	i_1	87	115	107	288	181	173	201	202	86
	j_1	9	12	11	31	19	18	21	21	8

3. A Deconstructive Sequence generated by the cycle A=0123456789.

Instead of sequentially repeating the digits 1-9 as in the case of the Smarandache Deconstructive Sequence we will use the digits 0-9 to form the corresponding sequence:

0,12,345,6789,01234,567890,1234567,89012345,678901234,678901234,56789012345,6
78901234567, ...

In this case the cycle has $n = 10$ elements. As we have seen in the introduction the sequence then has a period $= 2n = 20$. The periodicity starts for $i = 8$. Table 5 shows how for $i > 7$ any term C_i in the sequence is composed by concatenating a first part $B(k)$, a number q of cycles A="0123456789" and a last part $E(k)$, where $i = 7 + k + 20j$, $k = 1,2,...20$, $j \geq 0$, as expressed in (2) and $q = 2j, 2j + 1$ or $2j + 2$. In the analysis of the sequence it is important to distinguish between the cases where $E(k) = 0$, $k = 6,11,14,19$ and cases where $E(k)$ does not exist, i.e. $k = 8,12,13,14$. In order to cope with this problem we introduce a function $u(k)$ which will at the same time replace the functions $\delta(j)$ and $u=1+[\log_{10}E(k)]$ used previously. $u(k)$ is defined as shown in table 5. It is now possible to express C_i in a single formula

$$(8) \quad C_i = C_{7+k+20j} = E(k) + A * \sum_{r=0}^{q(k)+2j-1} (10^{10})^r + B(k) * (10^{10})^{q(k)+2j}) * 10^{u(k)} .$$

The formula for C_i was implemented modulus prime numbers less then 100. The result is shown in table 6 for $p \leq 41$. Again we note that the divisibility by a prime p is periodic with a period d which is equal to $p - 1$ or a divisor of $p - 1$, except for $p = 11$ and $p = 41$ which are factors of $10^{10} - 1$. The cases $p = 3$ and 5 have very simple answers and are not included in table 6.

Table 5

$n=10$, A$=0123456789$

i	k	B(k)	q	E(k)	u(k)
8+20j	1	89	2j	012345=3·5·823	6
9+20j	2	6789=3·31·73	2j	01234=2·617	5
10+20j	3	56789=109·521	2j	01234=2·617	5
11+20j	4	56789=109·521	2j	012345=3·5·823	6
12+20j	5	6789=3·31·73	2j	01234567=127·9721	8
13+20j	6	89	2j+1	0	1
14+20j	7	123456789=3²·3607·3803	2j	01234=2·617	5
15+20j	8	56789=109·521	2j+1		0
16+20j	9		2j+1	012345=3·5·823	6

17+20j	10	6789=3·31·73	2j+1	012=2^2·3		3
18+20j	11	3456789=3·7·97·1697	2j+1	0		1
19+20j	12	123456789=3^2·3607·3803	2j+1			0
20+20j	13		2j+2			0
21+20j	14		2j+2	0		1
22+20j	15	123456789=3^2·3607·3803	2j+1	012=2^2·3		3
23+20j	16	3456789=3·7·97·1697	2j+1	012345=3·5·823		6
24+20j	17	6789=3·31·73	2j+2			0
25+ 20j	18		2j+2	01234=2·617		5
26+20j	19	56789=109·521	2j+2	0		1
27+20j	20	123456789=3^2·3607·3803	2j+1	01234567=127·9721		8

Table 6

Divisibility of the 10-cycle destructive sequence by primes $7 \leq p \leq 41$

p=7	k	3	6	7	8	11	12	13	14	15	18	19	20
d=3	i_1	30	13	14	15	38	59	60	61	22	45	46	47
	j_1	1	0	0	0	1	2	2	2	0	1	1	1

p=11	k	1	2	3	4	5	6	7	8	9	10
d=11	i_1	88	9	110	211	132	133	74	35	176	137
	j_1	4	0	5	10	6	6	3	1	8	6
	k	11	12	13	14	15	16	17	18	19	20
	i_1	18	219	220	221	202	83	44	185	146	87
	j_1	0	10	10	10	9	3	1	8	6	3

p=13	k	2	3	4	12	13	14
d=3	i_1	49	30	11	59	60	61
	j_1	2	1	0	2	2	2

p=17	k	1	5	10	12	13	14	16
d=4	i_1	48	32	37	79	80	81	43
	j_1	2	1	1	3	3	3	1

p=19	k	1	2	3	4	5	10	12	13	14	16
d=9	i_1	128	149	90	31	52	117	179	180	181	63
	j_1	6	7	4	1	2	5	8	8	8	2

p=23	k	1	2	3	4	5	10	12	13	14	16
d=11	i_1	168	149	110	71	52	217	219	220	221	223
	j_1	8	7	5	3	2	10	10	10	10	10

p=29	k	2	4	10	12	13	14	16
d=7	i_1	129	11	97	139	140	141	43
	j_1	6	0	4	6	6	6	1

p=31	k	3	9	12	13	14	17
d=3	i_1	30	56	59	60	61	64
	j_1	1	2	2	2	2	2

p=37	k	2	3	4	12	13	14
d=3	i_1	9	30	51	59	60	61
	j_1	0	1	2	2	2	2

p=41	k	1	2	3	4	5	6	7	8	9	10
d=41	i_1	788	589	410	231	32	353	614	615	436	117
	j_1	39	29	20	11	1	17	30	30	21	5
	k	11	12	13	14	15	16	17	18	19	20
	i_1	678	819	820	821	142	703	384	205	206	467
	j_1	33	40	40	40	6	34	10	9	9	22

References

1. F. Smarandache, *Only Problems, Not Solutions*, Xiquan Publishing House, Phoenix, Arizona, 1993.

2. K. Kashihara, *Comments and Topics on Smarandache Notions and Problems,* Erhus University Press, Vail, Arizona, 1996.

3. C. Ashbacher, *Some Problems Concerning the Smarandache Deconstructive Sequence,* Journal of Recreational Mathematics, Vol 29, Number 2 – 1998, Baywood Publishing Company,Inc.

ZEROES IN THE DIGITS OF *N* FACTORIAL

Michael P. Cohen

1615 Q Street NW #T-1

Washington DC 20009-6310

mpcohen@juno.com

Abstract

We study the proportion of zero digits in the decimal (base 10) representation of $N!$, building on earlier work in the *Journal of Recreational Mathematics* by H. L. Nelson and Charles Ashbacher. The trailing zeroes (zeroes to the right of the rightmost nonzero digit) and internal zeroes (zeroes to the left of the trailing zeroes) are considered separately.

Introduction

In [1], Ashbacher investigates the proportion of zero digits in the decimal representation of $N!$, continuing earlier work by Nelson in [2]. We add to this investigation and answer a question posed by Ashbacher. In discussing the digits, we shall always be referring to the decimal (base 10) representation.

Trailing Zeroes

By a *trailing zero*, we mean a zero to the right of the rightmost nonzero digit. The number of trailing zeroes of $N!$ is determined by the number of 2's and 5's in its prime decomposition. It is easy to see there are always more 2's than 5's, so we concentrate on counting the number of 5's that divide $N!$. Each multiple of 5 less than or equal to N adds one, each multiple of 25 adds another one, each multiple of 125 adds yet another one, and so forth. Thus the number $T(N)$ of trailing zeroes in the decimal representation of $N!$ is

$$T(N) = \sum_{k=1}^{\infty} \left\lfloor \frac{N}{5^k} \right\rfloor$$

where $\lfloor \ \rfloor$ is the floor function, the greatest integer no larger than its argument. The quantity $\lfloor N/5^k \rfloor$ is zero when $5^k > N$, that is, when $k > \lfloor \log_5 N \rfloor$, so

$$T(N) = \sum_{k=1}^{\lfloor \log_5 N \rfloor} \left\lfloor \frac{N}{5^k} \right\rfloor.$$

I thought there was a chance this result might be new but, alas, it is prevalent on the Internet. See, for example, [3], [4], and [5].

The number of digits in $N!$ is $D(N) = \lfloor \log_{10} N! \rfloor + 1$, so the proportion of trailing zeroes is exactly $P(N) = T(N) / D(N)$. See Table 1 for the average value of $P(N)$ over ranges of a thousand at a time. These numbers agree with those of Ashbacher who did a direct count.

Table 1

The Mean Proportion of Trailing Zeroes in the Digits of N!

Range of N	P(N)	$\tilde{P}(N)$
1000-1999	0.090864	0.091556
2000-2999	0.084045	0.084666
3000-3999	0.080202	0.080443
4000-4999	0.077490	0.077695
5000-5999	0.075460	0.075637

Ashbacher asks how $P(N)$ behaves as N goes to infinity. To answer this question, first observe that

$$T(N) \le \sum_{k=1}^{\infty} \frac{N}{5^k} \le \left(\sum_{k=0}^{\infty} \frac{N}{5^k} \right) - N \le \frac{N}{1 - \frac{1}{5}} - N = N/4. \tag{1}$$

On the other hand,

$$T(N) \ge \sum_{k=1}^{\lfloor \log_5 N \rfloor} \left(\frac{N}{5^k} - 1 \right) = \left(\sum_{k=0}^{\lfloor \log_5 N \rfloor} \frac{N}{5^k} \right) - N - \lfloor \log_5 N \rfloor = N \frac{1 - \left(\frac{1}{5} \right)^{\lfloor \log_5 N \rfloor + 1}}{1 - \frac{1}{5}} - N - \lfloor \log_5 N \rfloor \tag{2}$$

$$= \frac{N}{4} - \frac{5N}{4} \left(\frac{1}{5} \right)^{\lfloor \log_5 N \rfloor + 1} - \lfloor \log_5 N \rfloor.$$

Hence $T(N)$ is asymptotic to $N/4$ by which we mean that their ratio converges to 1 as N goes to infinity. Hart, et al., in [6] also obtain the $N/4$ result but without the explicit non-asymptotic bounds provided by (1) and (2).

Looking now at $D(N)$, the number of digits in $N!$, we have $D(N) = \lfloor \log_{10} N! \rfloor + 1$. We apply a simple form of Stirling's formula [4] which asserts that $\ln N!$ is asymptotic to $N \ln N - N$ where $\ln$ is the natural logarithm. Then $D(N)$ is asymptotic to $\log_{10} N! = \ln N! / \ln 10$ which is asymptotic to $(N \ln N - N) / \ln 10$. Finally, the proportion $P(N)$ of trailing zero digits in the digits of $N!$ is $T(N) / D(N)$ which is asymptotic to

$$\tilde{P}(N) = \frac{N/4}{(N \ln N - N) / \ln 10} = \frac{(\ln 10)/4}{\ln N - 1}.$$

For example, $P(10,000) = 0.0700785\ldots$ and $\tilde{P}(10,000) = 0.0701123\ldots$. In answer to Ashbacher's question, $P(N)$ converges to 0 at a slow logarithmic rate.

Internal Zeroes

We call the digits to the left of the trailing zeroes *internal digits*. Clearly the leftmost and rightmost internal digits are nonzero. These digits we call *boundary digits*. The non-boundary internal digits may be zero. We find it necessary to make an assumption:

Assumption A: On average, one tenth of the non-boundary internal digits are zero.

If so, then the number of *internal zeroes* (non-trailing zeroes) is on average $[D(N) - T(N) - 2]/10$. The proportion of internal zeroes among all the digits is then on average

$$Q(N) = \frac{1}{10}\left(1 - \frac{T(N) + 2}{D(N)}\right).$$

$Q(N)$ is asymptotic to

$$\tilde{Q}(N) = \frac{1}{10}\left(1 - \frac{(\ln 10)/4}{\ln N - 1}\right).$$

In Table 2 we compare the mean values of $Q(N)$ and $\tilde{Q}(N)$ to the actual values obtained by Ashbacher.

Table 2

The Mean Proportion of Internal Zeroes in the Digits of $N!$

Under Assumption A

Range of N	Actual Mean from [1]	$Q(N)$	$\tilde{Q}(N)$
1000-1999	0.091023	0.090862	0.090844
2000-2999	0.091576	0.091568	0.091553
3000-3999	0.091836	0.091961	0.091956
4000-4999	0.092278	0.092237	0.092231
5000-5999	0.092402	0.092443	0.092436

Conclusion

In summary, we have obtained a rather complete description of the trailing zeros of the digits of $N!$. For internal zeroes we had to make an assumption.

Question: Is Assumption A valid? If not, what can replace it?

References

1. Charles Ashbacher, Some Results of an Investigation of Unsolved Problem #1079, *Journal of Recreational Mathematics, 35:1*, pp.1-4, 2006.

2. H. L. Nelson, Partial Solution to Problem 1079, *Journal of Recreational Mathematics, 15 :2*, pp 156-157, 1982-1983.

3. Elizabeth Stapel, Factorials and Trailing Zeroes, *Purple Math*, http://www.purplemath.com/modules/factzero.htm, last accessed April 8, 2015.

4. Eric Weisstein, Factorial, *Wolfram MathWorld*, http://www.mathworld.wolfram.com/Factorial.html, last accessed April 8, 2015.

5. Wikipedia Foundation, Trailing Zero, *Wikipedia*, http://en.wikipedia.org/wiki/Trailing_zero, last accessed April 8, 2015.

6. David S. Hart, James E. Marengo, Darren A. Narayan, and David S. Ross, On the Number of Trailing Zeros in n!, *The College Mathematics Journal, 39:2*, pp. 139-141, 2008.

THE MAGIC OF THREE

David L. Emory

137 Sycamore Lane

Lexington, VA 24450

emory@kalexres.kendal.org

Abstract

The meaning of period length is explained as a basis for understanding the significance of the number three. When an odd number is multiplied by three, a larger odd number is created with the same period length. This multiplication by three can take place again and again, but there is a limit. Here, only period lengths of six are examined, but many possibilities are demonstrated in a table.

When one (1.000…) is divided by any number not divisible by two or five, the result is a repeating decimal. The division eventually results in a remainder of one, at which point the repetition begins, since you divided into one in the first place. The number of digits in the quotient after the decimal point and before the repetition begins is the period length.

The following are two examples of the results of two such divisions:

$$1.000 / 77 = 0.012987012987$$

$$1.000 / 91 = 0.010989010989$$

The period length of 77 is six, and the period length of 91 is also six.

In calculating the period lengths of many odd numbers, I have found that nearly all period lengths are either a number one less than the odd number, or a simple fraction thereof. Simple fractions include one half (13, 31, 43, 47, 67, 71, and 89), one fourth (53), one eighth (41), and one ninth (73).

I knew, for example, that the period length of 23 is 22. In my research I had reached 69 and was totally surprised to discover its period length is also 22. The number one less than 69 is 68. Half of that is 34, and half of that is 17. Is 17 related to 22? I didn't think so. Finally I realized that 69 is three times 23.

As shown in Table 1, there are many examples where the multiplication by three of an odd number whose period length is known will produce a larger odd number with the same period length.

Table 1
The Period Lengths Are The Same

Odd Number	Period Length	Times Three	Period Length
7	6	21	6
11	2	33	2
13	6	39	6
17	16	51	16
19	18	57	18
21	6	63	6
23	22	69	22

29	28	87	28
31	15	93	15
41	5	123	5
43	21	129	21
47	23	141	23
49	42	147	42
53	13	159	13
59	58	177	58
67	33	201	33
69	22	207	22
71	35	213	35
73	8	219	8
89	44	267	44

   The numbers 7, 13, and 77 have period lengths of six. 7 times 13 is 91, which also has a period length of six. 13 times 77 is 1001, which has a period length of six (see Table 2), but I chose not to count numbers greater than three digits. Doing research for a completely different reason, I found that 259 has a period length of six.

   Earlier I discussed multiplication of odd numbers by three, which results in sixteen numbers less than one thousand that all have period lengths of six. They are: 7, 13, 21, 39, 63, 77, 91, 117, 189, 231, 259, 273, 351, 693, 777, and 819. I found it interesting that 7, 77, and 777 all have period lengths of six.

Table 2

Interesting Relationship of Period Lengths (P.L.)

N	$10^N + 1$	P. L. $= 2$ N
1	11	2
2	101	4
3	1001	6

4	10001	8
5	100001	10

Multiplication by three can take place again and again, but there are limits beyond which the period lengths are no longer six. Multiplying 7 by 3 four times makes 567, which has a period length of 18. Likewise, multiplying 13 by 3 four times makes 1053, which also has a period length of 18.

Having multiplied the two smallest numbers (7 and 13) by 3 four times, I thought I would do the same with the next smallest number (21), which was rather silly of me. I didn't stop to realize that 21 is already a multiple of three! Thus I multiplied 7 by 3 five times. This produces 1701 with a period length of 54.

The first three multiplications by 3 always create numbers with period lengths of 6, the fourth time with 18 and the fifth time with 54. Notice that 6 x 3 = 18 and 18 x 3 = 54.

Finally, I chose to find a divisor of one by multiplying seven by 3 six times and predicting the period length of the product (5103) to be 162. This turned out to be true.

Table 3

Multiplications of Seven by Threes

Divisor of One	Period Length
$7 \times 3^0 = 7 \times 1 = 7$	6
$7 \times 3^1 = 7 \times 3 = 21$	6
$7 \times 3^2 = 7 \times 9 = 63$	6
$7 \times 3^3 = 7 \times 27 = 189$	6
$7 \times 3^4 = 7 \times 81 = 567$	18
$7 \times 3^5 = 7 \times 243 = 1701$	54
$7 \times 3^6 = 7 \times 729 = 5103$	162

A BRIEF BIOGRAPHY OF AL-KASHI

Osama Ta'ani

Department of Mathematics
Plymouth State University
17 High Street, MSC 29
Plymouth, NH 03264
otaani@mail.plymouth.edu

Charles Ashbacher

5530 Kacena Ave
Marion, IA 52302
cashbacher@yahoo.com

Abstract

The man known simply as al-Kashi was born in Persia, now the modern nation of Iran, in the later part of the fourteenth century. He rose from a state of severe poverty to become an accomplished mathematician and scientist, working in many areas.

His textbook **Key to Arithmetic (Miftah al-Hisab)** and the abbreviated version **Concise Exposition of the Key** (**Talkhis Al-Miftah**) was used to teach mathematics for approximately two centuries in Persia and the Ottoman Empire. This paper is a brief description of his life along with some of his accomplishments.

The source for the material in this paper is "An Analysis of the Contents and Pegagogy of Al-Kashi's 1427 'Key to Arithmetic' (Miftah Al-Hisab)" by Osama Hekmatt Ta'ani, his Doctor of Philosophy dissertation at New Mexico State University.

The full name of the man known as al-Kashi was Ghiyath al-Din Jamshid bin Mas'ud bin Mahmood al-Tabeeb al-Kashi and he was born in the Persian (now Iranian) town of Kashan in 1380 CE. The name al-Kashi is the result of the common practice of the time of assigning a nickname based on the city of birth. Ghiyath al-Din is another nickname and means "the rescuer of religion." His first name was Jamshid, and the word "bin" means "son-of" and the names of his father and grandfather were Mas'ud and Mahmood respectively. Al-Tabeeb means "the physician," his profession at the start of his life in science. From this point on he will be referred to as al-Kashi.

Al-Kashi was born into severe poverty and was entirely self-taught, his first recorded activity in science was his observation of a lunar eclipse in 1406. One year later his first book, **The Ladder of Heavens of Solution for Difficulties Met by Forerunners in Determining Distances and Volumes** [of Celestial Bodies] was published.

Al-Kashi was very precise in his computations, in the introduction to his later work **Treatise on the Circumference (Al-Risala al-Muhityyah)** he states that with the current assumed size of the sphere of fixed stars of $6 * 10^5$ diameters of the Earth his goal is to compute the value of π so that the error would be smaller than the width of a single horse hair. His computation of π was done in sexagesimal (base 60) digits.

After the publication of his second book **Khaqani Zij**, al-Kashi was invited to the city of Samarqand (in present day Uzbekistan) by Ulugh Beg, an enlightened ruler that actively promoted learning and scholarship. Ulugh Beg asked al-Kashi to create an observatory in Samarqand that was larger than the one in Margha in Persia. Al-Kashi worked in that observatory until he died in 1429, leaving behind seven major works in mathematics and seventeen in astronomy. It was in 1427 when al-Kashi completed his compendium **Key to Arithmetic (Miftah al-Hiab)**.

Key to Arithmetic is composed of five sections: arithmetic of whole numbers, arithmetic of fractions, computations with sexagesimals, geometry and algebra. It was used as a mathematics textbook throughout the Islamic world to teach practical mathematics to a wide variety of professionals, from merchants and builders to astronomers and judges.

When the Ottoman Empire expanded out to its greatest extent, the **Key to Arithmetic** and the companion shortened version **Concise Exposition of the Key** were translated and used to teach mathematics throughout the vast lands of the empire.

Once the Ottoman Empire controlled the Arabian Peninsula in the sixteenth century, judges in Mecca studied **Key to Arithmetic** in order to learn the mathematics needed to solve inheritance problems. Subsequent mathematics textbooks published in the lands controlled by the empire also relied on the contents of **Key to Arithmetic** as well as the concise version.

Some historians refer to al-Kashi as the second Ptolemy, given the breadth of his writings as well as the impact they had on Middle Eastern societies for hundreds of years, this is clearly not an exaggeration.

Reference

O. H. Ta'ani, **An Analysis of the Contents and Pedagogy of Al-Kashi's 1427 "Key To Arithmetic" (Miftah Al –Hisab)**, New Mexico State University, 2011.

TRIANGULATION OF A TRIANGLE WITH TRIANGLES HAVING EQUAL INSCRIBED CIRCLES

Professor Ion Patrascu

Fraţii Buzeşti National College

Craiova, Romania

Professor Florentin Smarandache

University of New Mexico

Gallup, USA

Abstract

In this article, we solve the following problem:

Any triangle can be divided by a cevian into two triangles that have congruent inscribed circles.

We consider a given triangle ABC and we show that there is a point D on the side (BC) so that the inscribed circles in the triangles ABD, ACD are congruent. If ABC is an isosceles triangle $(AB = AC)$, where D is the middle of the base (BC), then the conclusion is immediate. Therefore, we assume that ABC is a non-isosceles triangle where D is a point on BC

We note I_1, I_2 the centers of the inscribed congruent circles; obviously, $I_1 I_2$ is parallel to the BC

(1). We observe that $m(\sphericalangle I_1 A I_2) = \frac{1}{2} m(\hat{A})$. (2)

Figure 1

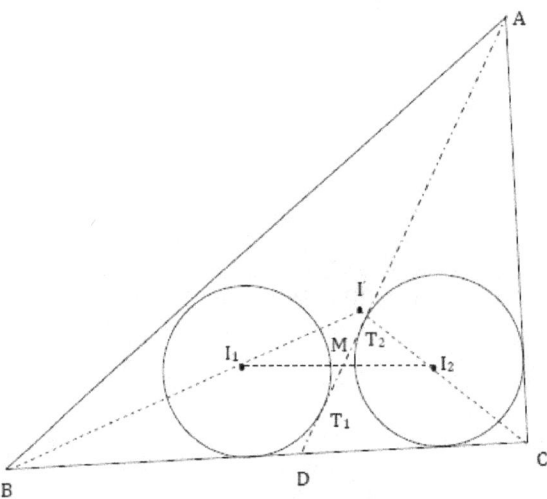

Let M be the intersection of $I_1 I_2$ with AD. If T_1, T_2 are contacts with the circles of the cevian AD, we have $\Delta I_1 T_1 M \equiv \Delta I_2 T_2 M$;

From this congruence, it is obvious that $(I_1 M) \equiv (I_2 M)$. (3)

Let I be the center of the circle inscribed in the triangle ABC; we will prove that: **AI is a simedian in the triangle $I_1 A I_2$.** (4)

Indeed, noting $\alpha = m(B\hat{A}I_1)$, it follows that $m(\sphericalangle I_1 AM) = \alpha$. From $\sphericalangle I_1 A I_2 = \sphericalangle BAI$, it follows that $\sphericalangle BAI_1 \equiv \sphericalangle IAI_2$, therefore $\sphericalangle I_1 AM \equiv \sphericalangle IAI_2$, indicating that AM and AI are isogonal cevians in the triangle $I_1 A I_2$. Since in this triangle AM is a median, it follows that AI is a bimedian.

Now, we show how we build the point D, using the conditions (1) – (4), and then we prove that this construction satisfies the stated requirements.

Building the Point D

1: We build the circumscribed circle of the given triangle ABC; we build the bisector of the angle BAC and denote by P its intersection with the circumscribed circle (see Fig. 2).

2: We build the perpendicular on C to CP and (BC) side mediator; we denote O_1 the intersection of these lines.

3: We build the circle $C(O_1; O_1C)$ and denote A' the intersection of this circle with the bisector AI (A' is on the same side of the line BC as A).

4: We build through A the parallel to $A'O_1$ and we denote it IO_1.

5: We build the circle $C(O_1'; O_1'A)$ and we denote I_1, I_2 its intersections with BI, and CI respectively.

Figure 2

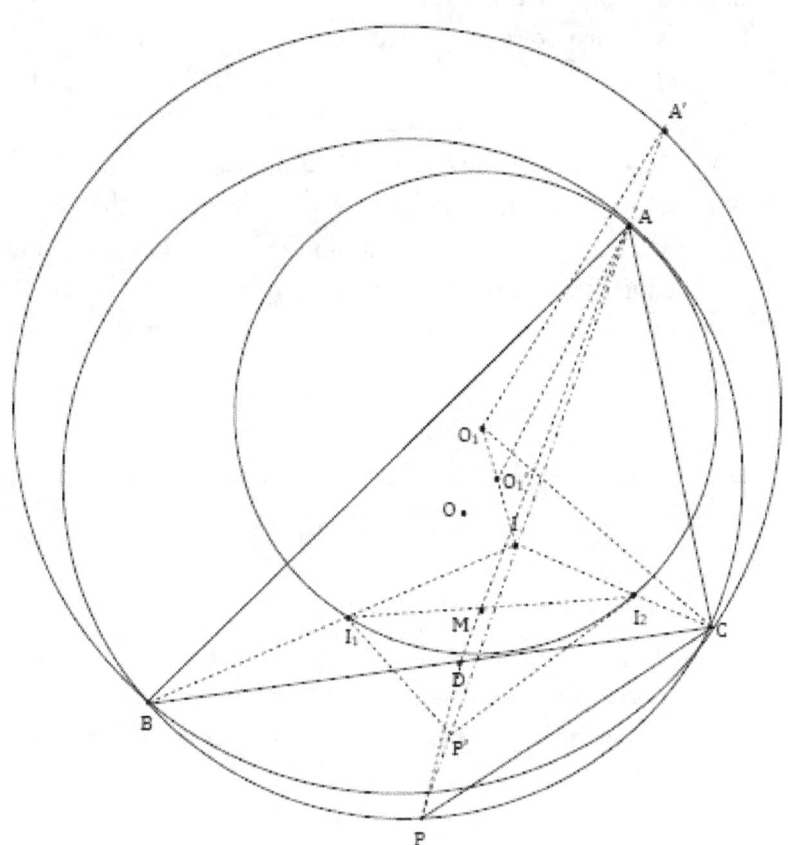

6: We build the middle M of the segment $(I_1 I_2)$ and denote by D the intersection of the lines AM and BC.

Proof

If the point P is the middle of the arc $\overset{\smile}{BC}$, then $m(\widehat{PCB}) = \frac{1}{2} m(\widehat{A})$.

The circle $C(O_1; O_1 C)$ contains the arc from which points the segment (BC) is shown under angle measurement $\frac{1}{2} m(\widehat{A})$.

The circle $C(O_1'; O_1' A)$ is homothetical to the circle $C(O_1; O_1 C)$ by the homothety of center I and by the report $\frac{IA'}{IA}$; therefore, it follows that $I_1 I_2$ will be parallel to the BC, and from the points of circle $C(O_1'; O_1' A)$ of the same side of BC as A, the segment $(I_1 I_2)$ „is shown" at an angle of measure $\frac{1}{2} m(\widehat{A})$. Since the tangents taken in B and C to the circle $C(O_1; O_1 C)$ intersect in P, on the bisector AI, as from a property of simedians, we get that $A'I$ is a simedian in the triangle $A'BC$. Due to the homothetical properties, it follows also that the tangents in the points I_1, I_2 to the circle $C(O_1'; O_1' A)$ intersect in a point P' located on AI, i.e. AP' contains the simedian (AS) of the triangle $I_1 A I_2$, noted $\{S\} = AP' \cap I_1 I_2$. In the triangle $I_1 A I_2$, AM is a median, and AS is simedian, therefore $\sphericalangle I_1 AM \equiv I_2 AS$; on the other hand, $\sphericalangle BAS \equiv \sphericalangle I_1 A I_2$; it follows that $\sphericalangle BAI_1 \equiv I_2 AS$, and more: $\sphericalangle I_1 AM \equiv BAI_1$, which shows that AI_1 is a bisector in the triangle BAD; plus, I_1, being located on the bisector of the angle B, it follows that this point is the center of the circle inscribed in the triangle BAD. Analogous considerations lead to the conclusion that I_2 is the center of the circle inscribed in the triangle ACD. Because $I_1 I_2$ is parallel to BC, it follows that the rays of the circles inscribed in the triangles ABD and ACD are equal.

Discussion

The circles $C(O_1; O_1 A')$, $C(O_1'; O_1' A)$ are unique; also, the triangle $I_1 A I_2$ is unique; therefore, the point D is unique.

Remark

At the beginning of the *Proof*, we assumed that ABC is a non-isosceles triangle with the stated property. There exists such triangles; we can construct such a triangle starting "backwards". We consider two given congruent external circles and, by tangent constructions, we highlight the ABC triangle.

Open problem

Given a scalene triangle ABC, could it be triangulated by the cevians AD, AE, with D, E belonging to (BC), so that the inscribed circles in the triangles ABD, DAE and the EAC to be congruent?

ALPHAMETICS

Edited by Charles Ashbacher

All of the alphametics in this section were created by Charles Ashbacher

The first point to note is the publication of the book **Alphametics Expressing Thoughts From the Star Trek™ Original Series**, written by Charles Ashbacher, ISBN 9781512152784.

The first problem is a tribute to the Star Trek original series episode 4, "The Naked Time" and the Star Trek: The Next Generation episode 3, "The Naked Now."

```
      3
      4
   BOTH    solve in base 14 and maximize NAKED
   TIME
    AND
    NOW
    ARE
  _____

  NAKED
```

This problem is a tribute to episode 6 of Star Trek: The Next Generation, "Where No One has Gone Before." It is a change from the orginal series, where the phrase was, "Where No Man Has Gone Before."

```
      6
  WHERE
  NOONE    solve in base 12 and minimize BEFORE
    HAS
   GONE
  _____

  BEFORE
```

A tribute to episode 23 of Star Trek: The Next Generation, "The Skin of Evil"

```
     23
   SKIN    solve in base 12 and maximize TASHA
     OF
   EVIL
  KILLS
  _____

  TASHA
```

This alphametic is doubly true in Adai, a Native American language in the Louisiana area of the United States. It went extinct in the early nineteenth century.

```
    NASS        2
    NASS        2
   COLLE        3
   COLLE        3
   _____      ___
  NEUSNE       10
```

Doubly true Somali

```
    EBAR    0
    AFAR    4    where AFAR is evenly divisible by 4
    AFAR    4
    LABO    2
   _____  ___
   TOBAN   10
```

Where two WRONGs can make a RIGHT. This problem is similar to one that appeared in **Mathematical Bafflers**, edited by Angela Dunn and published by Dover.

```
WRONG
  AND    and it is fitting that we want to maximize RIGHT
WRONG
_____
RIGHT
```

BOOK REVIEWS

Edited by:Charles Ashbacher

Charles Ashbacher Technologies

5530 Kacena Ave

Marion, IA 52302

E-mail: cashbacher@yahoo.com

College Calculus: A One-Term Course for Students with Previous Calculus Experience, by Michael E. Boardman and Roger B. Nelson, the Mathematical Association of America, Washington, D. C., 2015 388 pp., $60.00 (hardbound). ISBN 978-1-93951-206-2.

 When the authors use the phrase "with previous calculus experience" in the title they mean an introductory calculus course in high school. Using this as justification, they start late in the calculus sequence and move quickly.

 The best way to demonstrate how late they start is to point out that using substitution in integration is mentioned on page 7 and the explanation of the cylindrical shell method of determining volume starts on page 11. Integration by parts is on page 18.

 The material covered is typical of that of a second semester calculus course, but there are only brief encounters with the topics, theorems are stated but there is little in the way of proofs. After the techniques of integration are examined, numerical integration, polar coordinates, improper integrals and infinite series are covered.

 One characteristic of this book that makes it so unlike other calculus books is the small number of exercises. While there are exercises at the end of sections, the numbers and page count is unusually small. For example, the topic of section 8.3 is polar coordinates and there are only 26 exercises that take up approximately one-and-a-half pages at the end. I found this very refreshing, a welcome break from the apparent arms race where calculus book authors seem to try to differentiate themselves by including more exercises. Standard practice is followed in terms of solutions, answers to the odd-numbered ones are given in an appendix.

 If your course is one where you are presenting the essentials of calculus with minimal proof to students with a background in the derivative and have integration experience, this is a book that will work for you. Outside that audience, I am uncertain if the value extends beyond the person that needs a fast and furious refresher.

Charles Ashbacher

Mathematical Bafflers, edited by Angela Dunn, Dover Publications, Mineola, New York, 1980. 217 pp., $10.95 (paper). ISBN 9780486239613.

This review was published on the Mathematical Association of America Book Review site

http://www.maa.org/publications/maa-reviews/mathematical-bafflers

and is republished here with permission.

The world of mathematical puzzles is a rich one, yet is generally based on a small number of fundamental principles. Even though you may know the solution strategy for a type of puzzle, however, when presented with a new one you are essentially starting over. For example, the logic type of puzzle (John, George, Sam and Joe are married to Joan, Jenna, Janet and Judy. John is not married to Judy…) is a staple and many of that type are found in this book. Solving the problem means determining what you know and following a sequence of scenarios until the solution is found. No matter how many you solve, if the puzzle is well written it is still a challenge.

To me, the quality of the writing of the puzzles is what differentiates this book from the others. Having read many puzzle books in my years as a reviewer, my identification of the type of problem is often immediate. While that did often happen, there were many where I had to slow down my reading pace so that I could clearly understand the scenario.

The puzzles in this book were selected from the problems that appeared in the weekly "Problematical Recreations" column in *Aviation Week* and *Electronic News*. These are considered the best that appeared over the course of 12 years. Very little in the way of mathematical skill is needed to solve them, yet that does not make them easy, even for professional mathematicians/problem solvers. All solutions are included and like the great puzzles, generally obvious after the fact.

Math teachers from late elementary school all the way through college will be able to find something in this book that they can use to enliven their classrooms. No calculators or other computational devices are required, only an open brain and perhaps pencil and paper.

Charles Ashbacher

Exploring Advanced Euclidean Geometry with GeoGebra, by Gerard A Venema, the Mathematical Association of America, Washington, D. C., 2013. 129 pp, $50.00 (hardbound). ISBN 978-0-88385-784-7

Geometry is such a visual subject that it cannot be learned without the creation of understandable diagrams. In the classrooms of past years, this meant a writing utensil on either a chalk or whiteboard, a process that was not easy for those of us that are challenged by simple drawings. The combination of this text and the ability to use a free package that can be used to illustrate the examples is a powerful tool for teaching Euclidean geometry.

The speed with which the geometry can be taught as well as the level of student retention will both be increased by using this book as a text in combination with Geogebra. Since not much has changed in the area of advanced Euclidean geometry in a long time, the content is fairly standard.

I was very impressed with the last chapter where the Poincaré disk is used to demonstrate hyperbolic geometry. There is a misplaced notion among many that non-Euclidean geometry is hard when in fact there are very understandable ways to introduce it. This shows a logical and effective way to demonstrate hyperbolic geometry.

Utilizing the dynamic and visual advantages of Geogebra, the math teacher can delve far deeper into the subject matter than was possible before. Since the text is very clear as a standalone tool, the combination will make the teaching of geometry far more efficient than it has been in the past.

Charles Ashbacher

The Moscow Puzzles: 359 Mathematical Recreations, by Boris A. Kordemsky, edited by Martin Gardner, Dover Publications, Mineola, New York. 320 pp., $14.95 (paper). ISBN 978-0486270784.

This review was published on the Mathematical Association of America Book Review site

http://www.maa.org/publications/maa-reviews/the-moscow-puzzles-359-mathematical-recreations

and is republished here with permission.

Most of the puzzles in this collection have appeared in some form in many other publications, both in print and online. For example, number 11 is the classic, "Wolf, Goat and Cabbage" problem that can be traced back to writings in the eighth century. There are cryptarithms, designs with matches, dissections, logic problems in textual form, problems with dominoes, number crossword puzzles, puzzles involving magic squares, number puzzles and properties and a few involving chess and checkers.

Although the puzzles are generally old and have been frequently used, that is a tribute to their quality rather than an indication that they are stale and out of date. Instructors from elementary school all the way through college will be able to find items that they can use in their classes to challenge the students.

When I was in the sixth grade the math teacher posed a series of puzzles to the class and it was a competitive contest to solve them as the names of the solvers were posted on the board. All were within the capabilities of the class and we enjoyed the challenge and the thrill of solution. Some of the problems in this collection could have been used in that contest.

Study after study has demonstrated that the older person that continues to pursue mental challenges remains much more functional in the cognitive sense than the person that simply goes mentally passive. While it of course cannot solve all problems of losing mental acuity as you age, there are enough challenges in this book to help keep the neurons firing at a high level.

Charles Ashbacher

Headstrong: 52 Women Who Changed Science—and the World, by Rachel Swaby, Broadway Books, New York, New York, 2015. 288. pp., $16.00(paper). ISBN 978-0553446791.

Marie Curie. That is the name you'd probably hear if you walked up to a stranger and asked them to name a female scientist. But what would you hear if you pressed them to name another? Yvonne Brill? Emmy Noether? Salome Gluecksohn Waelsch? All of these women—Curie included—made significant contributions to their various fields. Yet most are strangers, their achievements buried under more familiar names like Einstein or Newton, or worse, buried under the stereotypes of their gender. **Headstrong** brings these women forward, profiling 52 female chemists, biologists, physicists, geneticists, and others whose ideas, research, dedication, and breakthrough discoveries paved the way for scientific advancement today.

Headstrong is broken up into seven sections—medicine, biology and the environment, genetics and development, physics, the earth and the stars, math and technology, and invention—with at least four women profiled in each category. Each profile is about four pages and briefly covers each woman's background, how she came to be interested in her work, and the significant contributions she made to her field. Swaby livened up what would otherwise be tedious writing with humorous quotes or situations taken directly from the lives of these women, a greatly appreciated personal touch.

While this book would be a great starting point for students who wanted to learn or write about women in science, as a casual read, it left something to be desired. I went into this book recognizing only five of the 52 women by name, but I left with only a handful more, feeling somewhat overwhelmed—not necessarily by the number of women profiled, but by the brevity with which they were presented. I wanted to know more! Swaby does include notes and a bibliography at the end of the book, useful for further reading, but I wish such information had also been included at the end of each woman's profile.

In spite of its brevity, I love that this book exists. In a time when women still feel left out of or discriminated within STEM fields, **Headstrong** offers 52 fantastic female role models—like Hedy Lamarr, whose contributions led to wi-fi, you can purchase every volume of **Topics in Recreational Mathematics** online while sitting at your local coffee shop. Thanks to these 52 women (and the many more not included in this book), anyone can be inspired to pursue an interest in science and perhaps make their own lasting contribution to the world and how we live in it.

Rachel Pollari

SOLUTIONS TO PROBLEMS THAT APPEARED IN JOURNAL OF RECREATIONAL MATHEMATICS 38(1)

Edited by Lamarr Widmer
Messiah College,
Box 3041
One College Avenue,
Grantham, OA 17027

widmer@messiah.edu

*2870. Prime Conjecture by Andrew Cusumano, Great Neck, NY (*JRM, 38*:1, p. 60)

Let p_n denote the n-th prime and N be the next-prime function, i.e. $N(m)$ is the smallest prime greater than m. Define $F(k) = N(2 \times 3 \times \ldots \times p_{k-1} + p_k) - (2 \times 3 \times \ldots \times p_{k-1})$. For example,

$$F(4) = N(2 \times 3 \times 5 + 7) - (2 \times 3 \times 5) = 11.$$

What can be said about the range of this function? Does it consist only of primes? Does it contain all odd primes?

Solution by Richard Hess

The data in table 1 suggests that only primes will be produced by the function

$$F(k) = N\left(\prod_1^{k-1} p_i + p_k\right) - \prod_1^{k-1} p_i$$

Since $F(k)$ is the difference between a prime larger that $|\ \prod_1^{k-1} p_i$ and $\prod_1^{k-1} p_i$, it is clear that $F(k)$ must be larger than p_{k-1}. So $F(k)$ has an infinite range.

For $F(k)$ to be composite we need a prime gap of at least p_{k-1}^2 after $\prod_1^{k-1} p_i$. The table shows that there are no such cases in the first sixteen entries. The closest candidate is $F(34) = 643$, but $p_{33} = 137$ and 137^2 is much larger than 643. It seems very unlikely that $F(k)$ will ever be composite.

Many odd primes do not belong to the range of $F(k)$. This is because the set of $F(k)$ for $k < n$ does not include p_{n-1}. Some such primes are 3, 17, 19, 31, 43, 53, 59, 79, 83, 97 and 109.

Table 1

n	N	F(n)	n	F(n)	n	F(n)	n	F(n)
2	7	5	17	73	32	163	47	773
3	13	7	18	89	33	229	48	607
4	41	11	19	109	34	643	49	383
5	223	13	20	89	35	239	50	383
6	2333	23	21	103	36	157		
7	30059	29	22	163	37	167		
8	510551	41	23	151	38	439		
9	9699727	37	24	197	39	239		
10	223092907	37	25	101	40	199		
11	6969693291	61	26	103	41	191		
12	200560490197	67	27	233	42	199		
13	7420738134871	61	28	223	43	383		
14	304250263527281	71	29	127	44	233		
15	13082761331670097	67	30	223	45	751		
16	614889782588491517	107	31	191	46	313		

2871. Pentomino Doublers by Lamarr Widmer, Mechanicsburg, PA (*JRM, 38*:1, p. 61)

Figure 1, a solution to problem 2822 (*JRM 36*:4, p. 359), shows that the *Z*-pentomino and three others can be used to construct a double scale version of itself. We readily see that the *X*-pentomino does not share this property. What about the other ten pentominoes (Figure 2)?

Figure 1

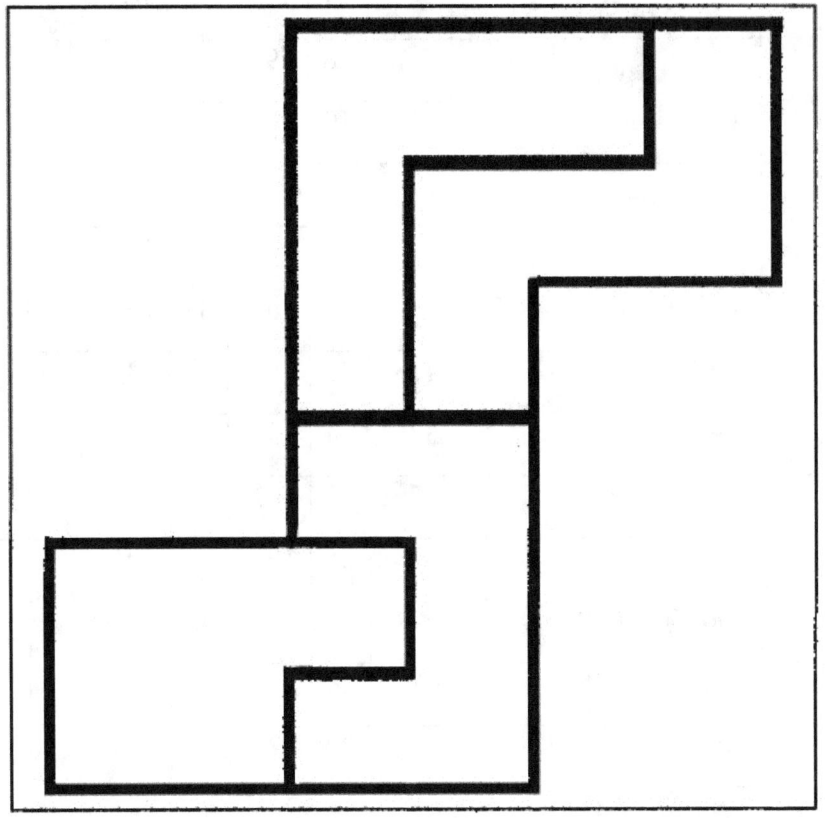

Figure 2

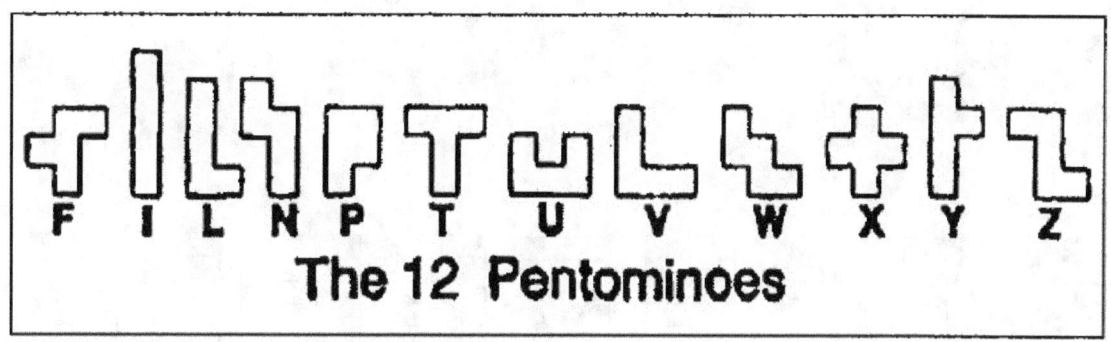

The 12 Pentominoes

Solution by Henry Ibstedt, Richard Hess and Brian Barwell (composite)

Seven pentominoes F, I, L, N, P, U and Z lent themselves to form the 33 diagrams below. FPUV, LPUX, NPVZ and PUVY gave raise to doubles of the doublers. It is remarkable that P occurs in all the diagrams except #31 LNVZ.

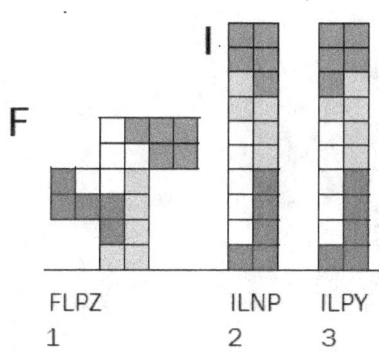

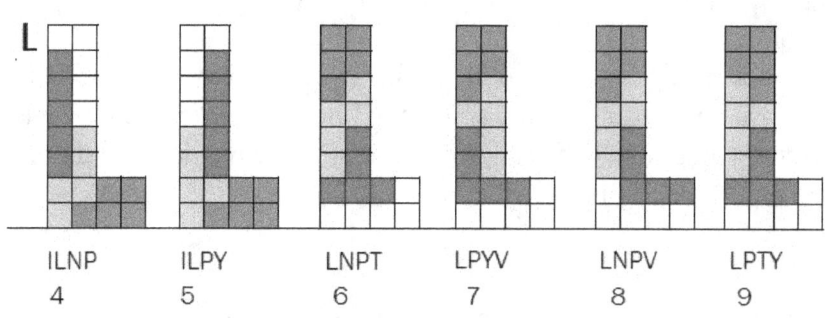

127

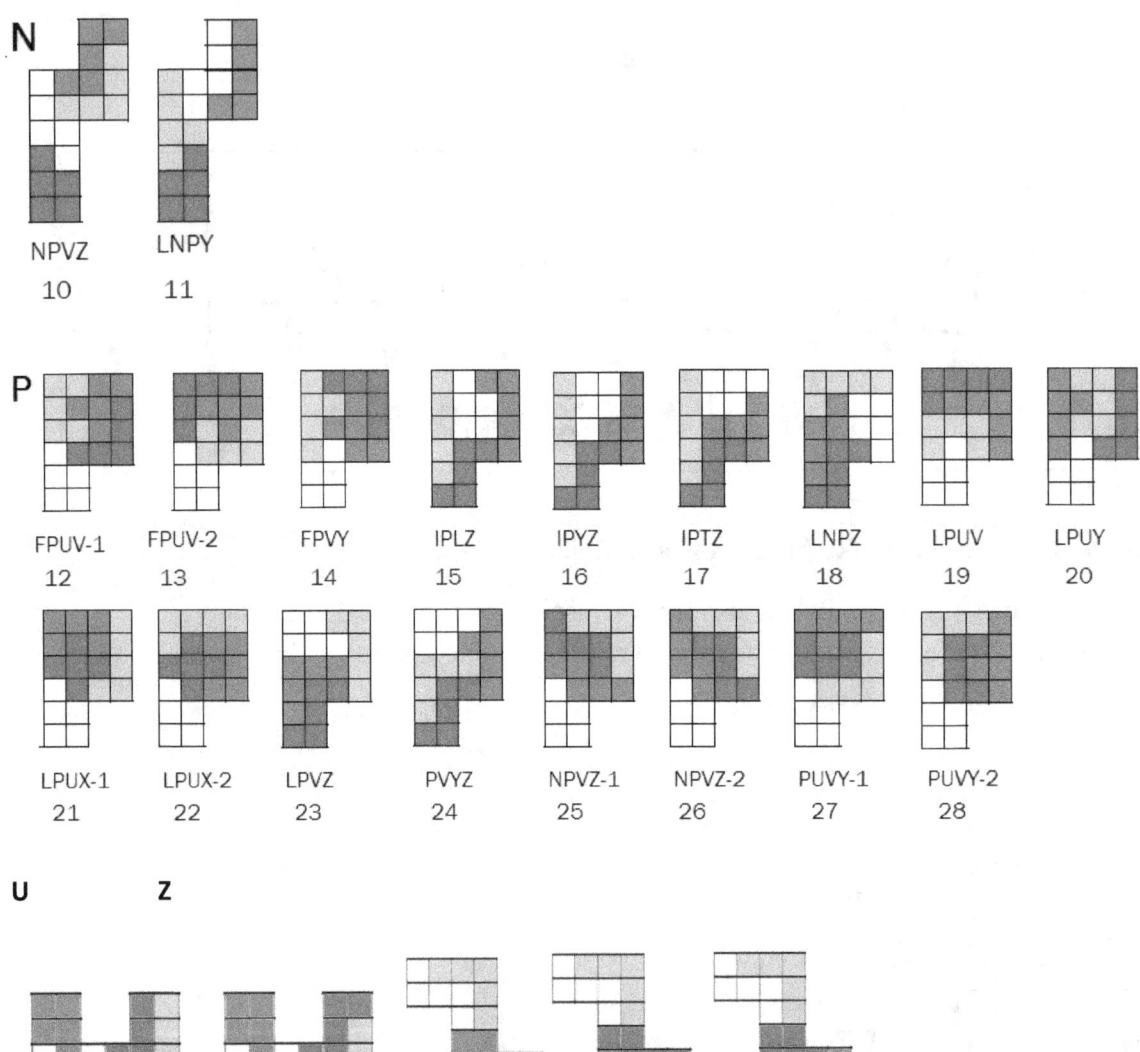

2872. Pipe Navigating a Corner by Hubert Hagadorn, Menlo Park, CA (*JRM, 38*:1, p. 61)

A pipe is bent up to ninety degrees at its midpoint so as to maximize its length for transporting in trenches of constant width having right angle turns. What is the maximum length of the pipe and what is the angle of the bend? Assume that the pipe remains horizontal while passing along the trench and that the diameter of the pipe is negligible.

Solution by the Proposer

The angle of bend is equal to 2α, *or* 47.071 degrees. The pipe length is $2L$, or 5.009 units. For reference, the angle of θ is 35.352 degrees when the pipe clearance is zero is (i.e. when the pipe just contacts the inside corner).

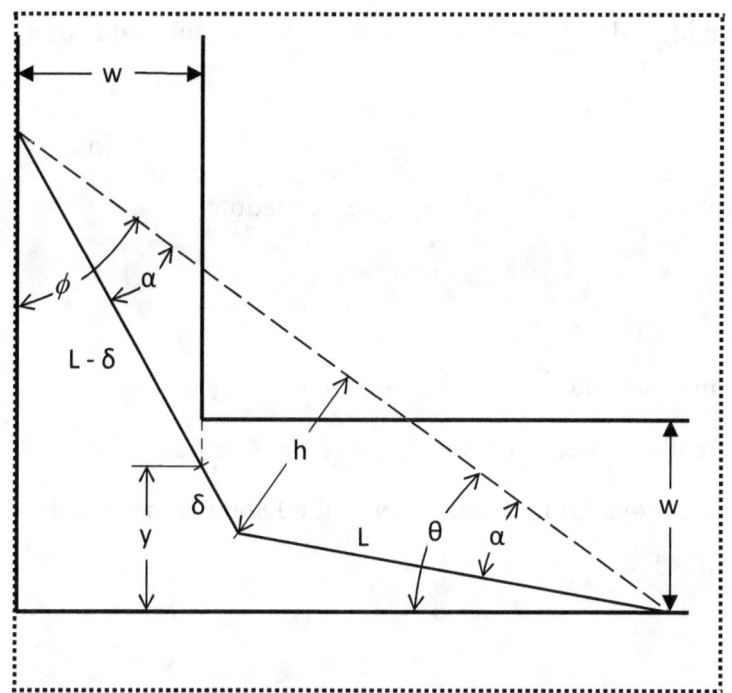

In the above figure the clearance of the pipe from the inside corner must be positive, requiring $y \le w$. The pipe length is $2L$, the trench width is w, while the pipe orientation is given by θ, which references a line passing through the pipe's endpoints to the horizontal.

$$y = L\sin(\theta - \alpha) + \delta\cos(\phi - \alpha),$$

where

$$\phi = \pi/2 - \theta$$

$$\delta = L - \frac{w}{\sin(\phi - \alpha)}$$

Eliminating δ and ϕ from y, and simplifying

$$y = 2L\sin(\theta)\cos(\alpha) - w\tan(\theta + \alpha) \tag{1}$$

Now h must not exceed w, so that

$$h = L\sin(\alpha) = kw, \ 0 < k \le 1 \tag{2}$$

Solving for L and substituting in (1)

$$y = w\,[2k\sin(\theta)\cot(\alpha) - \tan(\theta + \alpha)] \tag{3}$$

As the pipe negotiates a right turn, θ varies from 0 to 90 degrees. As a result y reaches a maximum then decreases. Finding the local maximum, and thus setting $dy/d\theta = 0$ gives

$$2k\cos(\theta)\cot(\alpha) - \sec^2(\theta + \alpha) = 0 \tag{4}$$

Since L is maximized when y in (3) is maximized, setting y to its maximum allowable value of w gives

$$2k \sin(\theta) \cot(\alpha) - \tan(\theta + \alpha) = 1 \qquad (5)$$

For a given k the above two equations may be solved for α and θ. From equation (2)

$$\frac{L}{w} = \frac{k}{\sin(\alpha)}.$$

Experimentally, L was found to be maximum when k is at its greatest value, unity.

2873. Unknown Modulus by Hubert Hagadorn, Menlo Park, CA (*JRM, 38*:1, p. 62)

Given nonnegative integers x and r, find solutions for the modulus n of the Diophantine equation $x \equiv r \pmod{n}$. Assume that $0 \le r \le n - 1$.

Solution by the Proposer

1. If $r > x$, no solution exists for n.

2. If $r = x$, n is any integer greater than r.

3. If $x / 2 \le r < x$, no solution exists for n.

4. If $0 \le r < x / 2$, n is any integer that is a divisor of $x - r$ and greater than r.

Condition 1 is a result that the modulus function yields r values that are less than or equal to x.

Condition 2 holds since for $n > x$, r will always be equal to x.

For Condition 3, consider the following equation upon which the modulus function is based

$$x = k n + r,$$

where k is the least positive integer greater than or equal to zero, and $0 \le r < n$. When k is zero solutions for $r = x$ applies. For $k \ge 1$, solving for n and requiring $n > r$

$$n = (x - r) / k > r,$$

so that

$$r < x / (k + 1).$$

Maximum values of r must be less than $x / 2$, corresponding to $k = 1$. Larger values of k will be more restrictive. As a result n has no solution for $x / 2 \le r < x$ (Condition 3).

For $r < x / 2$, possible values for k are the divisors of $x - r$. Possible values for n are likewise the divisors of $x - r$. The allowed values for n are those divisors greater than r (Condition 4).

130

For example, consider $x = 29$ and $r = 15$. No solution is possible as $r > x/2$. 29 mod 15 = 14, and 29 mod 14 = 1, not 15.

Also, consider $x = 29$ and $r = 5$. The divisors of $x - r = 24$ are 1, 2, 3, 4, 6, 8, 12, and 24. Valid values for n are the divisors greater than 5, namely, 6, 8, 12, and 24.

2874. Digital Fraction Sum by Andy Pepperdine, Bath, UK (*JRM, 38*:1, p. 62)

We can express the number 1 in several ways as the sum of two fractions which together use each of the nine nonzero digits exactly once. The fractions are not necessarily in lowest terms.

For example, $1 = \dfrac{4}{12} + \dfrac{638}{957}$.

a. What is the smallest possible value for any of the fractions?

b. If we include zero, but not as the first digit in any term, then what is the smallest possible value for one of the fractions?

Solution by several contributors

a. The smallest fraction is equal to $\dfrac{1}{19}$ as in $\dfrac{3}{57} + \dfrac{864}{912} = 1$.

b. The smallest fraction is equal to $\dfrac{1}{89}$ as in $\dfrac{4}{356} + \dfrac{792}{801} = 1 = \dfrac{6}{534} + \dfrac{792}{801}$.

2875. Integral Powers of an Irrational Number by Henry Ibstedt, Broby, Sweden (*JRM, 38*:1, p. 62)

Show that every positive integral power of $\sqrt{2} - 1$ can be expressed in the form $\sqrt{m} - \sqrt{m-1}$.

Solution by Michael P. Cohen

Define two sequences a_k and b_k by $a_1 = b_1 = 1$ and for $k > 1$, $a_k = a_{k-1} + 2b_{k-1}$ and $b_k = a_{k-1} + b_{k-1}$. We shall show that

$$(\sqrt{2} - 1)^k = \sqrt{m_k} - \sqrt{m_k - 1}$$

where if k is odd, $m_k = 2b^2_k$ and $m_k - 1 = a_k^2$. Whereas if k is even, $m_k = a_k^2$ and $m_k - 1 = 2b^2_k$.

The result holds for $k = 1$ because

$$(\sqrt{2} - 1)^1 = \sqrt{2b_1^2} - \sqrt{a_1^2}.$$

The result holds for $k = 2$ because

$$(\sqrt{2} - 1)^2 = (2 - 2\sqrt{2} + 1) = \sqrt{9} - \sqrt{8} = \sqrt{a_2^2} - \sqrt{2b_2^2}$$

where $a_2 = a_1 + 2b_1 = 3$ and $b_2 = a_1 + b_1 = 2$.

To prove the result for $k > 1$ odd by induction, we assume it holds for $k - 1$ even. Then

$$(\sqrt{2} - 1)^k = (\sqrt{2} - 1)^{k-1}(\sqrt{2} - 1) = (\sqrt{a_{k-1}^2} - \sqrt{2b_{k-1}^2})(\sqrt{2} - 1) = \sqrt{2a_{k-1}^2} - \sqrt{a_{k-1}^2} - 2\sqrt{b_{k-1}^2} + \sqrt{2b_{k-1}^2}$$

$$= \sqrt{2}a_{k-1} - a_{k-1} - 2b_{k-1} + \sqrt{2}b_{k-1} = \sqrt{2}(a_{k-1} + b_{k-1}) - (a_{k-1} + 2b_{k-1}) = \sqrt{2}b_k - a_k = \sqrt{2b_k^2} - \sqrt{a_k^2}.$$

Setting $m_k = 2b_k^2$, it remains to show that $a_k^2 = m_k - 1$. But

$$2b_k^2 - a_k^2 = 2(a_{k-1} + b_{k-1})^2 - (a_{k-1} + 2b_{k-1})^2 = 2a_{k-1}^2 + 4a_{k-1}b_{k-1} + 2b_{k-1}^2 - a_{k-1}^2 - 4a_{k-1}b_{k-1} - 4b_{k-1}^2$$

$$= a_{k-1}^2 - 2b_{k-1}^2 = 1$$

by the induction hypothesis.

Finally, we show the result for even $k + 1$. Similar to the previous equation,

$$(\sqrt{2} - 1)^{k+1} = (\sqrt{2} - 1)^k (\sqrt{2} - 1) = (\sqrt{2b_k^2} - \sqrt{a_k^2})(\sqrt{2} - 1) = 2\sqrt{b_k^2} - \sqrt{2a_k^2} - \sqrt{2b_k^2} + \sqrt{a_k^2}$$

$$= 2b_k - \sqrt{2}a_k - \sqrt{2}b_k + a_k = (a_k + 2b_k) - \sqrt{2}(a_k + b_k) = a_{k+1} - \sqrt{2}b_{k+1} = \sqrt{a_{k+1}^2} - \sqrt{2b_{k+1}^2}.$$

To show that $2b_{k+1}^2 = a_{k+1}^2 - 1$, compute

$$2b_{k+1}^2 - a_{k+1}^2 = 2(a_k + b_k)^2 - (a_k + 2b_k)^2 = 2a_k^2 + 4a_kb_k + 2b_k^2 - a_k^2 - 4a_kb_k - 4b_k^2$$

$$= a_k^2 - 2b_k^2 = -1.$$

2876. Dudeney's Century Puzzle by Andy Pepperdine, Bath, UK (*JRM, 38*:1, p. 62)

H.E. Dudeney in his *Amusements in Mathematics*, question 90, "The Century Puzzle", asks us to represent the number 100 as a "mixed fraction", that is of the form $A + B/C$, using each of the digits, $1 - 9$, once only. For example $100 = 82 + 3546 / 197$. He says that Edouard Lucas had found seven solutions, but in fact there are 11, and in that Dudeney was correct. There are 11 solutions, which he supplies in his answers.

a. What about other powers of ten? Can the numbers 10, 1000, 10000, etc. be so represented, and if so, in how many ways?

b. What about using all ten digits, not allowing zero as the first digit in any number?

c. In Question 91, "More Mixed Fractions", he states in his answers that, for the number 26, he had "recorded no fewer than 25 different arrangements" using the nine digits. In fact there are more. How many?

Solution by the Proposer and by Hubert Hagadorn (independently)

For (a) and (b) all representations are included in table 2.

Table 2

	Nine digits	Ten digits
10	6 + 5892 / 1473	1 + 85203 / 9467
	7 + 5469 / 1823	6 + 19032 / 4758
	7 + 5496 / 1832	6 + 30192 / 7548
	7 + 6549 / 2183	6 + 37140 / 9285
	7 + 6954 / 2318	
	7 + 9546 / 3182	
	7 + 9654 / 3218	
100	3 + 69258 / 714	27 + 65043 / 891
	81 + 5643 / 297	36 + 57024 / 891
	81 + 7524 / 396	43 + 51072 / 896
	82 + 3546 / 197	45 + 21780 / 396
	91 + 5742 / 638	51 + 34692 / 708
	91 + 5823 / 647	72 + 13860 / 495
	91 + 7524 / 836	73 + 24516 / 908
	94 + 1578 / 263	82 + 10674 / 593
	96 + 1428 / 357	
	96 + 1752 / 438	
	96 + 2148 / 537	

	Nine digits	**Ten digits**
1000	534 + 9786 / 21	153 + 60984 / 72
	597 + 4836 / 12	208 + 41976 / 53
	597 + 8463 / 21	396 + 27180 / 45
	751 + 9462 / 38	561 + 40827 / 93
	756 + 4392 / 18	745 + 21930 / 86
	913 + 4872 / 56	843 + 15072 / 96
	924 + 3876 / 51	957 + 4386 / 102
	951 + 4263 / 87	957 + 8643 / 201
	954 + 3726 / 81	964 + 3852 / 107
	957 + 3612 / 84	987 + 4056 / 312
	967 + 1254 / 38	

	Nine digits	Ten digits
10000	348 + 57912 / 6	1047 + 26859 / 3
	451 + 76392 / 8	3691 + 50472 / 8
	631 + 74952 / 8	4785 + 31290 / 6
	948 + 27156 / 3	4908 + 15276 / 3
	978 + 54132 / 6	5041 + 39672 / 8
	7914 + 6258 / 3	5401 + 36792 / 8
	9316 + 5472 / 8	5491 + 36072 / 8
	9541 + 3672 / 8	7503 + 14982 / 6
	9753 + 1482 / 6	7845 + 12930 / 6
		9345 + 7860 / 12
		9435 + 6780 / 12
		9637 + 5082 / 14
		9702 + 5364 / 18
		9745 + 8160 / 32
		9765 + 4230 / 18
		9853 + 6027 / 41
100000	None	321 + 598074 / 6
		376 + 498120 / 5
		483 + 597102 / 6
		651 + 298047 / 3

For (c), there are 29 ways of representing 26 using the nine non-zero digits:

3 + 21758 / 946	18 + 4736 / 592	21 + 3485 / 697
4 + 16258 / 739	18 + 5392 / 674	21 + 3845 / 769
8 + 17352 / 964	18 + 5432 / 679	21 + 4685 / 937

135

9 + 12546 / 738	18 + 5936 / 742	21 + 4835 / 967
12 + 6398 / 457	18 + 6352 / 794	21 + 4865 / 973
14 + 3576 / 298	18 + 7456 / 932	23 + 1974 / 658
18 + 3672 / 459	18 + 7536 / 942	24 + 1358 / 679
18 + 3752 / 469	18 + 7624 / 953	24 + 1538 / 769
18 + 4296 / 537	18 + 7632 / 954	24 + 1586 / 793
18 + 4632 / 579	19 + 5236 / 748	

And 30 ways if we include a zero:

3 + 24587 / 1069	8 + 57042 / 3169	9 + 54706 / 3218
6 + 34580 / 1729	9 + 26078 / 1534	9 + 58072 / 3416
6 + 35840 / 1792	9 + 34867 / 2051	9 + 68357 / 4021
6 + 38540 / 1927	9 + 35768 / 2104	9 + 68527 / 4031
6 + 43580 / 2179	9 + 46801 / 2753	9 + 71536 / 4208
6 + 47180 / 2359	9 + 47651 / 2803	9 + 78251 / 4603
6 + 54380 / 2719	9 + 51476 / 3028	9 + 80512 / 4736
6 + 58340 / 2917	9 + 51782 / 3046	9 + 86241 / 5073
6 + 71840 / 3592	9 + 52768 / 3104	9 + 86734 / 5102
7 + 30286 / 1594	9 + 54026 / 3178	24 + 6158 / 3079

2877. Recurring 3's by Charles Ashbacher, Marion, Iowa (*JRM, 38*:1, p. 63)

In his book *Mathematics Galore*, James Tanton poses the problem of finding a number N such that all of the multiples $N, 2N, ..., 10N$ contain a digit 3.

a. Tanton provides the solution $N = 19507893$ which "more than" solves this problem. Find the first positive integer m, such that mN does **not** contain a digit 3.

b. Find a value of N which improves on the one given in a. That is, find N, such that all of the multiples $N, 2N, ..., kN$ contain a digit 3 and $k \geq m$.

*c. Is there an upper bound on the value of k?

Reference:

1. *Mathematics Galore,* James Tanton, The Mathematical Association of America, Washington D.C., 2012. ISBN 978-0-88385-776-2.

Solution by Michael P. Cohen

a. By direct computation, m = 26. That is, all multiples $N, 2N, \ldots, 25N$ contain a digit 3 but $26N$ does not.

b. We will construct an N such that k is at least 1,000. Every number of at most 1,000 when multiplied by at least one of the numbers 1, 2, 3, 4, 5, 6, 7, 8, 9, 11, 12, 13,14,15, 16, 17, 25, or 31 contains a digit 3 and has at most 5 digits. Take

N=31000250001700016000150001400013000120001100009000080000700006000050000400003000020000l.

This number will have a digit of 3 when multiplied by any number 1,000 or less. The zeroes between the positive digits prevent carries from interfering.

c. No. The construction method given in b can be extended, adding more zeroes between the positive digits, to give arbitrarily long sequences.

Editor's Commentary

Andy Pepperdine answered part c with the following.

There is no limit to the value of k.

First, note that for each value j, we can choose a number $v(j)$ such that $j \times v(j)$ contains the digit 3. To see this consider that all single digits have a multiple between 30 and 39, all two digit numbers have a multiple that lies between 300 and 399, and all three digit numbers a multiple that lies between 3000 and 3999, etc.

Now look at the numbers $v(1), v(2), v(3), \ldots v(k)$, and put them in sequence one after the other, and place sufficient zeroes between them to remove any carries when multiplied by each j in turn. The resulting number, when multiplied by j, will contain a three in the j'th section.

2878. Primes of the Form $p - 2^k$, by Henry Ibstedt, Broby Sweden (*JRM, 38*:1, p. 63)

a. For the prime $p = 3331$, the value of $p - 2^k$ is prime for odd values of k from 1 through 11. Is there a longer such sequence of primes?

b. For the prime $p = 1487$, the value of $p - 2^k$ is prime for even values of k from 2 through 10. Is there a longer such sequence of primes?

Solution by Hubert Hagadorn

a. There are many longer sequences of such primes. Based on searches to 10^8, table 3 contains the smallest primes found for an odd k > 11.

Table 3

Prime	k
754939	13
5308579	15
10786879	17
190475011	21

b. There are many longer sequences of such primes. Based on searches to 10^8, table 4 contains the smallest primes found for an even k > 10.

Table 4

Prime	k
29867	12
49433	14
176417	16
13032533	20

2879. Linear Combination of Irrationals, by Hubert Hagadorn, Menlo Park, CA (*JRM, 38*:1, p. 63)

We consider the inequality $|a\pi + be + c\varphi| < 10^{-6}$ where π, e and φ (golden ratio) are mathematical constants, while a, b and c are integers. Find solutions where $abc \neq 0$, and

a. $a + b + c = 0$

b. $|a| + |b| + |c|$ is minimum.

Solution by Andy Pepperdine

Assuming that the golden ratio is defined as $\varphi = (\sqrt{5} + 1)/2 \approx 1.618034$.

a) If $a + b + c = 0$, then $c = -(a + b)$, and we are looking for $|a(\pi - \varphi) + b(e - \varphi)| < 10^{-6}$

We need a value of a for which $|a\frac{\pi - \varphi}{e - \varphi}|$ is close to an integer. A computer search turns up

$a = 751809$, $b = -1041061$, $c = 289252$, and $a\pi + be + c\varphi \approx -8.286443 \times 10^{-7}$.

b) A computer search finds:

$a = 229$, $b = -1173$, $c = 1526$, $a\pi + be + c\varphi \approx -2.780580 \times 10^{-7}$, and $|a| + |b| + |c| = 2928$.

138

Editor's Commentary

Hubert Hagadorn notes that part (a) may readily be solved by evaluating continued fractions of the ratio of $\frac{b}{a} = \frac{-(e-\varphi)}{\pi-\varphi}$. The twelfth continued fraction gives the ratio $-1041061/751809$, satisfying the accuracy criterion. Hence $a = 751809$, $b = -1041061$, and $= -(a + b)$.

PROPOSERS AND SOLVERS LIST FOR PROBLEMS AND CONJECTURES THAT APPEARED IN JOURNAL OF RECREATIONAL MATHEMATICS, 38(1)

P	S	Name	Location	28 70	28 71	28 72	28 73	28 74	28 75	28 76	28 77	28 78	28 79
■		Charles	Marion, IA								P		
	■	Michael P. Cohen	Washington,				S		S		S		S
■		Andrew Cusumano	Great Neck, New York	P									
■	■	Hubert Hagadorn	Menlo Park, CA			P	P	S	S	S	S	S	P
	■	Richard I. Hess	Rancho Palos Verdes, CA	S	S	S	S	S	S	S		S	S
■	■	Henry Ibstedt	Broby, Sweden		S				P		P		
		Ken Klinger	Northbrook, IL	S								S	S
■	■	Andy Pepperdine	Bath, UK		S			P	S	P	S		S
■	■	Proposer/Solver	P = Proposer S = Solver										

SOLUTIONS TO THE ALPHAMETICS IN THIS ISSUE

Charles Ashbacher

1.

```
                    3
                    4
     11   7  13     3    where 0 & 3 and 6 & 8 can interchange
     13   4   8     9
         12   1     5
          1   7     0
         12   6     9
     _____
  1  12  10   9     5
```

2.

```
                    6
     4   8   0   3  0
     7   2   2   7  0
             8   9  6
        11   2   7  0
     _____
  1  0  10   2   3  0
```

3.

```
              2   3
        9  10  6   1
              5   0    where 0 & 1 can interchange
        4   2   6   7
    10  6   7   7   9
    _____
    11  8   9   3   8
```

4.

```
       1933
       1933
      68224
      68224
    _____
     140314
```

5.

```
  9764
  6164   where 8 & 9 can interchange
  6164
  8670
  _____

 30762
```

6.

```
 49217
   316
 49217
 _____

 98750
```

Edgematching puzzles, the neighborly mathematics

by Kate Jones

The idea of linking compatible parts together is as old as the horse and cart, or a handshake, or dominoes. Early forms of matching puzzles are jigsaw-cut pictures where pieces join to re-establish the original image. Those pieces, however, fit only one way.

A more interesting puzzle allows any piece to match with any other piece on one or more sides. The pioneering work in this genre was done by British mathematician Major Percy MacMahon, who published several models in his landmark book, *More Mathematical Pastimes* (1921). The most famous are MacMahon's Three-Color Squares and MacMahon's Four-Color Triangles.

 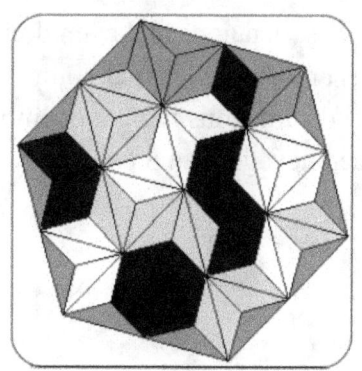

The Three-Color Squares are divided diagonally into four triangles which are individually colored so that each side of the square can have its own color. By coloring these subtriangles in all different combinations of three colors, MacMahon identified 24 distinct tiles. These can then be joined into various shapes where all touching sides match and the outer border is uniform. The 4x6 rectangle has 13,328 solutions! Much of the research into this set was done by Wade Philpott in the 1960s and 1970s and published first by Wade and later by Kadon under the name *Multimatch I*. Philpott identified every solvable symmetrical shape, from shortest to longest perimeter.

The Four-Color Triangles consist of equilateral triangles divided into three isosceles subtriangles to give each side of the triangle its own color. When filled with every combination of four colors, the resultant 24 triangles can be joined in many shapes, the most compact being an order-2 hexagon (left), with all edges matched and a uniform border. Wade Philpott's extensive research identified every possible symmetrical shape solvable with both conditions. This set has been published by Kadon since 1989 as *Multimatch III*.

These two sets begat the design of many related themes of edgematching tiles with various shapes and connection methods. All product names shown are trademarks of Kadon Enterprises, Inc. All sets not credited by name to others are by Kate Jones. Here are some examples from Kadon's growing collection, left to right: *Multimatch II* (corner-colored squares); *Multimatch IV* (corner-colored triangles); and *Marshall Squares* (edge-colored squares with 1 or 2 colors per tile, using 5 colors) by William Rex Marshall.

 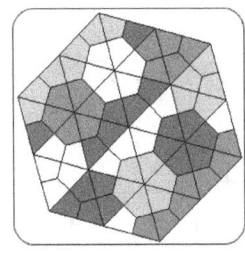 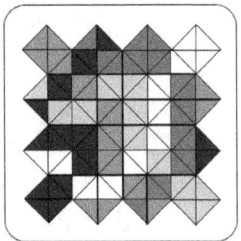

More elaborate sets use hexagons, octagons and even dodecagons, left to right: *Hexmozaix* (hexagons inlaid with chevrons and diamonds in 3 colors) by Charles Butler; *MemorIQ* (hexagons inlaid with 3 flattened pentagons in every combination of 4 colors) and *Doris* (octagons inlaid with 4 stretched hexagons in 3 colors) by Zdravko Zivkovic.

A most majestic set of 30 tiles is *Dazzle* (below left) invented by Charles Butler, where dodecagons are inlaid with pie-shaped segments in 3 colors that match joined tiles while they surround triangular spaces with all same or all different colors. *Grand Bowties* (below right) has 4 colors inlaid on cross-shaped tiles that can match on edges, corners or even just tips.

Another connecting device uses cut-outs that can even be assembled by touch to enable blind players to solve, like these, left to right: *Four on a Match* (squares with 4 shapes embedded at their corners); *MultiTouch I* (squares with 3 shapes carved from their edges—equivalent to Multimatch I); *MultiTouch II* (squares with 3 shapes cut into their corners—equivalent to Multimatch II); *MultiTouch III* (triangles with 4 shapes on their edges—equivalent to

144

Multimatch III); and *MultiTouch IV* (triangles with 4 shapes carved from their corners—equivalent to Multimatch IV). When matched cut-outs join, interesting filigree patterns appear.

Changing colors or cut-outs into contour shapes also produces equivalent sets of matchable tiles, like these, left to right: *Snowflake Super Square* (squares with 3 shapes of edge instead of color, related to Multimatch I); *Trifolia* (triangles with 4 shapes of edge substituted for colors—equivalent to Multimatch III); and *Leaves* (hexagons with inward and outward notches) created by Sjaak Griffioen.

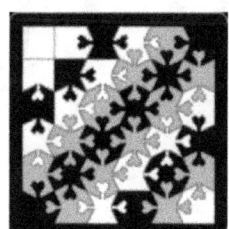 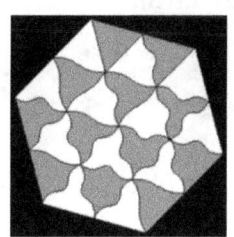 

Further innovations are tiles with pathways on them that produce continuous lines and loops when joined. From left to right, top to bottom: *Arc Angles* (25 distinct curved kites with line segments connecting five edge points); *Dezign-8* (squares with 1 through 4 path exits); *Kaliko*

(classic set of 85 hexagons with 3 colors on 5 path configurations that link opposite sides) created by Titus and Schensted; and *Fractured Fives* (two-sided squares with paths on one side and dissected pentominoes on the other).

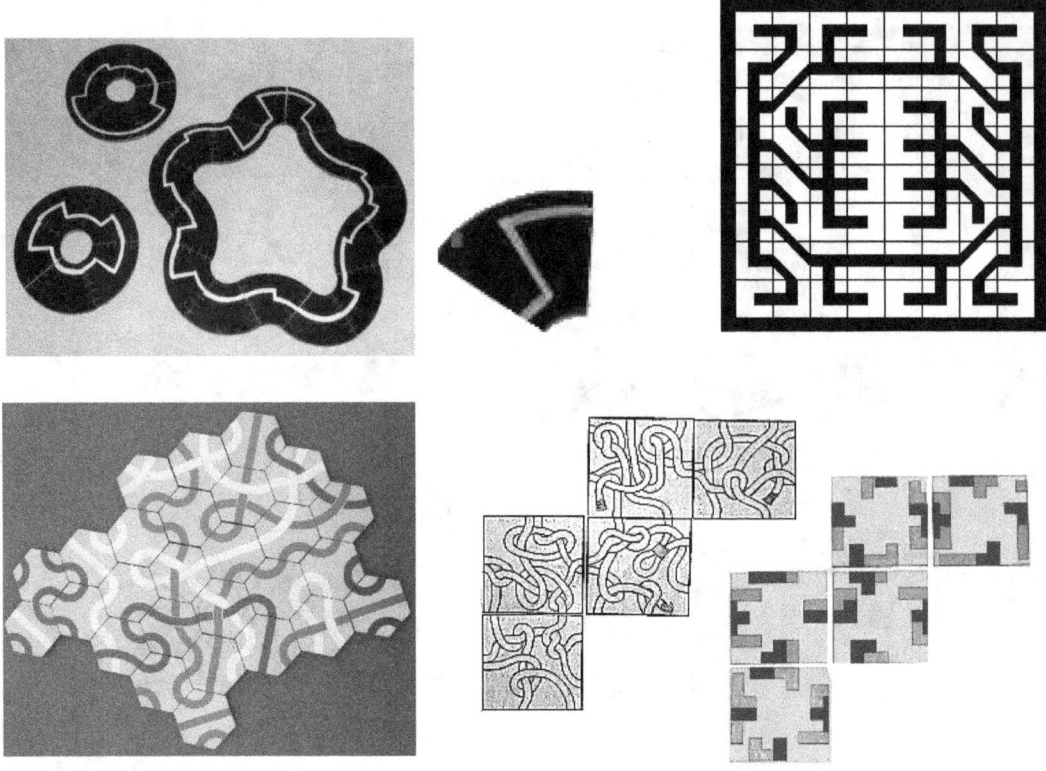

These puzzles are eloquent paradigms of systems that can build harmony from diversity in a great many cooperative ways. There's hope that human societies could do likewise! You can see these puzzles and more, most of them suitable for ages 8 to adult, on Kadon's website at **www.gamepuzzles.com/edgemtch.htm**

Mathematical Spectrum

*A magazine for students
and teachers of mathematics
in schools, colleges and universities*

Editor: D. W. Sharpe, *University of Sheffield*

For over three decades, *Mathematical Spectrum* has been a popular source of stimulating ideas for teachers, students and mathematical enthusiasts alike. Articles cover a wide range of topics in mathematics and the related sciences as well as the history of mathematics, with regular education and computer columns, a letters page, problems and solutions, and reviews of books and software.

Contributors from all over the world include established mathematicians as well as students — we welcome original student contributions and award annual prizes for the best ones published.

Subscription information: Vol. 47 (Sept 14–Aug 15) $24.50; Vols 47 and 48 (Sept 14–Aug 16) $47.00; Vols 47, 48 and 49 (Sept 14–Aug 17) $67.50. Three issues per volume in September, January and May; postage and handling included. To subscribe, contact:

Mathematical Spectrum
Applied Probability Trust
School of Mathematics and Statistics
The University of Sheffield
Sheffield S3 7RH, UK

Tel: +44 114 222 3922
Fax: +44 114 222 3926

Email: s.c.boyles@shef.ac.uk
Web: www.appliedprobability.org

Published by the **Applied Probability Trust**, a non-profit-making organisation based in the University of Sheffield

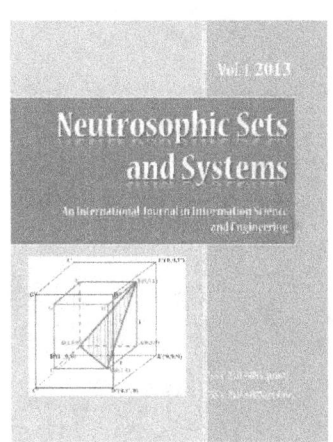

Neutrosophic Sets and Systems has been created for publications on advanced studies in neutrosophy, neutrosophic set, neutrosophic logic, neutrosophic probability, neutrosophic statistics that started in 1995 and their applications in any field, such as the neutrosophic structures developed in algebra, geometry, topology, etc.
The submitted papers should be professional, in good English, containing a brief review of a problem and obtained results. Neutrosophy is a new branch of philosophy that studies the origin,

nature, and scope of neutralities, as well as their interactions with different ideational spectra.
This theory considers every notion or idea <A> together with its opposite or negation <antiA> and with their spectrum of neutralities <neutA> in between them (i.e. notions or ideas supporting neither <A> nor <antiA>). The <neutA> and <antiA> ideas together are referred to as <nonA>.
Neutrosophic Set and Logic are generalizations of the fuzzy set and respectively fuzzy logic (especially of intuitionistic fuzzy set and respectively intuitionistic fuzzy logic). In neutrosophic logic a proposition has a degree of truth (T), a degree of indeter
minacy (I), and a degree of falsity (F), where T, I, F are standard or non-standard subsets of $]^-0,$ $1^+[$.
Neutrosophic Probability is a generalization of the classical probability and imprecise probability.
Neutrosophic Statistics is a generalization of the classical statistics.
What distinguishes the neutrosophics from other fields is the <neutA>, which means neither <A> nor <antiA>.
<neutA>, which of course depends on <A>, can be indeterminacy, neutrality, tie game, unknown, contradiction, ignorance, imprecision, etc.

All submissions should be designed in MS Word format using our template file:

   http://fs.gallup.unm.edu/NSS/NSS-paper-template.doc

A variety of scientific books in many languages can be downloaded freely from the Digital Library of Science:

   http://fs.gallup.unm.edu/eBooks-otherformats.htm

To submit a paper, mail the file to the Editor-in-Chief. To order printed issues, contact the Editor-in-Chief. This journal is non-commercial, academic edition. It is printed from private donations.

Information about the neutrosophics you get from the UNM website:
   http://fs.gallup.unm.edu/neutrosophy.htm
The home page of the journal is accessed on
   http://fs.gallup.unm.edu/NSS

Topics in Recreational Mathematics 1/2015

Presenting papers and articles in recreational mathematics or material of interest to people interested in recreational mathematics. Original artwork with a mathematical theme will also be featured.

Contents

Editor-in-chief
Charles Ashbacher

Assistant editor
Rachel Pollari

Artwork
Caytie Ribble

Technical assistant
Gisela Hausmann

Dedicated to the legacy of Martin Gardner and Joseph S. Madachy

Available on Amazon

ISBN 978-1507603215

Topics in Recreational Mathematics
2/2015

Presenting papers and articles in recreational mathematics or material of interest to people interested in recreational mathematics. Original artwork with a mathematical theme is also featured.

Editor-in-Chief: Charles Ashbacher
Assistant editor: Rachel Pollari
Artwork: Caytie Ribble
Technical advisor: Gisela Hausmann

Contents

Available on Amazon   ISBN 978-1508617099

TOPICS IN RECREATIONAL MATHEMATICS 3/2015

Presenting papers and articles in recreational mathematics or material of interest to people interested in recreational mathematics. Original artwork with a mathematical theme is also featured.

Contents

Available on Amazon ISBN: 9781511641005

ALPHAMETICS AS EXPRESSED IN RECREATIONAL MATHEMATICS MAGAZINE

Alphametics have been a staple of recreational mathematics since the first issue of **Recreational Mathematics Magazine**. A column of alphametics appeared in the first issue of RMM and it was a regular feature in **Journal of Recreational Mathematics** throughout the 38 ½ volumes that it was published.

This book contains the alphametics and their solutions that appeared in **Recreational Mathematics Magazine** during the 14 issues that it was published by Joseph S. Madachy.

Contents

Editor's Notes
by Charles Ashbacher

Mathematical Cartoon
by Caytie Ribble

Introduction
by Charles Ashbacher

Solving Addition Alphametics
by Charles Ashbacher

The Alphametics That Appeared in Recreational Mathematics Magazine

Solutions to Alphametics

Available on Amazon

ISBN 978-1508538134

Editor-in-chief
Charles Ashbacher

Artwork
Caytie Ribble

TEN YEAR CUMULATIVE INDEX TO THE JOURNAL OF RECREATIONAL MATHEMATICS

Edited by

Joseph S. Madachy

Updated by Charles Ashbacher

This is a republication of the Ten Year Index published by Baywood Publishing Company in 1982.

Available on Amazon, ISBN 9781508936800

Alphametics Expressing Thoughts
From the Star Trek Original Series

```
        WHERE
        NOMAN
          HAS
         GONE
       -------
       BEFORE
```

Authored by Charles Ashbacher

Illustrated by Caytie Ribble and Jenna Richardson

 This is a collection of alphametics, one for each of the episodes of the Star Trek™ original series as well as the first pilot. Solutions to all of the problems are also included.

 Three cartoons having a Star Trek theme and an original image of the late Leonard Nimoy as Spock are also included. The cartoons were drawn by Caytie Ribble and the image of Spock by Jenna Richardson.

Available on Amazon ISBN: 9781512152784

Mathematical Cartoons

Ideas by Charles Ashbacher

Cartoons drawn by Caytie Ribble

This is a collection of 50 cartoons having a mathematical theme. The ideas and concepts expressed range from the vulgar (in the mathematical sense) to the asinine, literally and figuratively. Brief explanations of the cartoons are also included.

Available on Amazon ISBN: 9781514207130